THE COMPLETE GUIDE TO
SPACE

An Imprint of Sterling Publishing
1166 Avenue of The Americas
New York, NY, 10036

Text © 2015 by QEB Publishing, Inc.
Illustrations © 2015 by QEB Publishing, Inc.

This 2015 edition published by Sandy Creek.

ISBN: 978-1-4351-6165-8

Manufactured in China
Lot #:
2 4 6 8 10 9 7 5 3 1
08/15

THE COMPLETE GUIDE TO
SPACE

AMANDA ASKEW

Sandy Creek
NEW YORK

CONTENTS

>> THE UNIVERSE

SPACE IS EVERYWHERE 6

THE BIG BANG 8

HOW FAR? 10

FOCUS ON... ALBERT EINSTEIN 12

WHAT ARE GALAXIES? 14

ALL SHAPES AND SIZES 16

CRASH! BANG! 18

BRIGHTEST GALAXIES 20

THE MILKY WAY 22

FOCUS ON... EDWIN HUBBLE 24

>> THE STARS

WHAT ARE STARS? 26

THE SUN 28

THE SUN'S FEATURES 30

SOLAR ECLIPSES............................. 32

STAR BIRTH 34

LARGE AND SMALL........................ 36

DEATH OF A STAR 38

BLACK HOLES 40

>> THE SOLAR SYSTEM

WHAT IS THE SOLAR SYSTEM? 42

THE SOLAR SYSTEM BEGINS..................... 44

CIRCLING THE SUN............................. 46

FOCUS ON... JOHANNES KEPLER 48

FOCUS ON... ISAAC NEWTON 50

THE ROCKY PLANETS 52

MERCURY, THE SMALLEST PLANET............................ 54

VENUS, THE HOTHOUSE 56

EARTH, THE LIVING PLANET 58

INSIDE EARTH 60

SPINNING EARTH 62

THE MOON 64

CHANGING SHAPES OF THE MOON 66

MARS, THE RED PLANET 68

Words in **bold** are explained in the Glossary on page 140.

THE GIANT PLANETS 70

JUPITER, THE LARGEST PLANET 72

FOCUS ON... GALILEO GALILEI............................ 74

SATURN, THE RINGED PLANET 76

URANUS, THE TILTED PLANET 78

FOCUS ON... WILLIAM HERSCHEL 80

NEPTUNE, THE WINDIEST PLANET 82

THE DWARF PLANETS 84

PLUTO, THE DWARF PLANET 86

ASTEROIDS, THE LEFTOVER ROCKS 88

DIRTY SNOWBALLS 90

FAMOUS COMETS 92

METEORS AND METEORITES 94

>> EXPLORING SPACE

WHAT IS SPACE EXPLORATION? 96

EARLY EXPLORERS 98

ROCKETING INTO SPACE 100

THE SPACE SHUTTLE 102

ASTRONAUTS... 104

SPACE STATIONS.. 106

INSIDE THE ISS 108

EYE IN THE SKY 110

MISSIONS TO THE MOON 112

WALKING ON THE MOON 114

VISITING THE ROCKY PLANETS 116

EXPLORING MARS 118

VISITING THE GIANTS 120

JOURNEY TO THE SUN 122

>> ASTRONOMY

WHAT IS ASTRONOMY? 124

STAR MAPS 126

GAZING AT THE STARS 128

HOW TO USE STAR MAPS 130

STARS IN THE NORTHERN HEMISPHERE .. 132

STARS IN THE SOUTHERN HEMISPHERE 134

TELESCOPES AND OBSERVATORIES 136

TELESCOPES IN THE SKY........................... 138

GLOSSARY 140

INDEX 143

SPACE IS EVERYWHERE

The **Universe** is huge and includes everything in **space**, from large **stars** and **galaxies** to tiny pieces of dust. It stretches for billions of miles—much farther than we could ever imagine or explore. The Universe was created billions of years ago in an explosion called the Big Bang.

STAR FACT

Space begins about 60 miles above sea level. That's the same as 215 Empire State Buildings piled on top of one another!

There are billions of galaxies in space, of many different colors and shapes. Each galaxy is made up of billions of stars—just like the Milky Way.

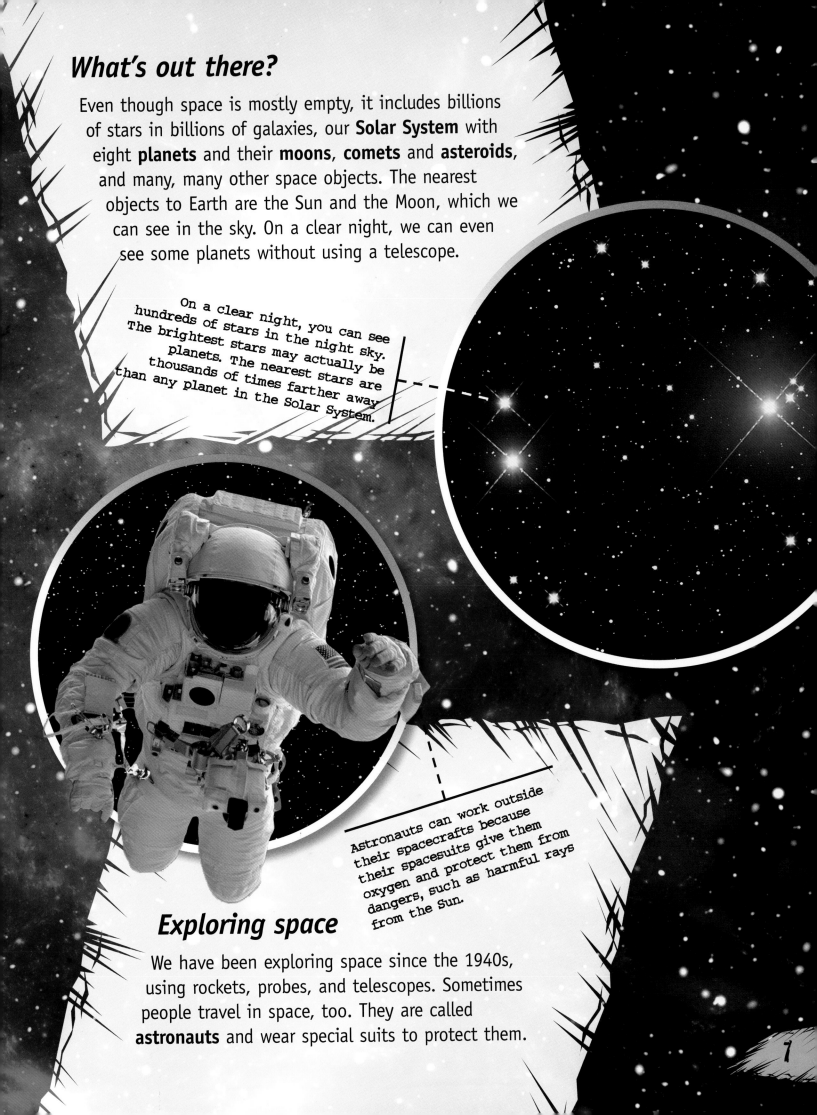

What's out there?

Even though space is mostly empty, it includes billions of stars in billions of galaxies, our **Solar System** with eight **planets** and their **moons**, **comets** and **asteroids**, and many, many other space objects. The nearest objects to Earth are the Sun and the Moon, which we can see in the sky. On a clear night, we can even see some planets without using a telescope.

On a clear night, you can see hundreds of stars in the night sky. The brightest stars may actually be planets. The nearest stars are thousands of times farther away than any planet in the Solar System.

Astronauts can work outside their spacecrafts because their spacesuits give them oxygen and protect them from dangers, such as harmful rays from the Sun.

Exploring space

We have been exploring space since the 1940s, using rockets, probes, and telescopes. Sometimes people travel in space, too. They are called **astronauts** and wear special suits to protect them.

THE BIG BANG

Scientists believe that the Universe began about 14 billion years ago in a massive explosion called the Big Bang. The explosion sent everything racing outward, and the Universe quickly started to spread out and has been getting larger ever since.

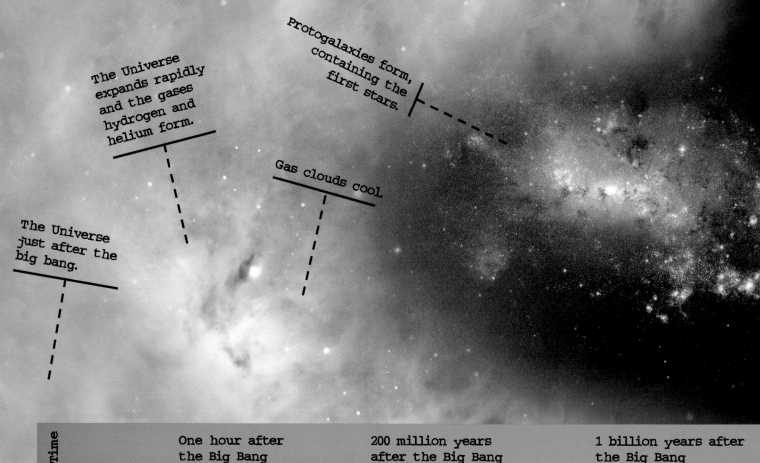

The Universe has changed over billions of years, forming new galaxies of different shapes and sizes.

Protogalaxies form, containing the first stars.

The Universe expands rapidly and the gases hydrogen and helium form.

Gas clouds cool

The Universe just after the big bang.

| Time | One hour after the Big Bang | 200 million years after the Big Bang | 1 billion years after the Big Bang |

Creating the Universe

At first, the Universe was very small and very hot. As this heat **energy** started to spread out and rush in every direction, it began to cool. As it cooled, the energy turned into **matter**. Matter makes up everything in the Universe—every **solid**, **liquid**, and **gas**.

8

The cooling energy formed clouds of helium and hydrogen gas. The gas clouds collected into clumps where stars began to form. Some of these clumps merged together to create galaxies. This is still happening today.

Large galaxies form—some become spirals.

Protogalaxies start to merge to form galaxies.

3 billion years after the Big Bang

5 billion years after the Big Bang

Pulling together

The gas clouds were pulled together by a force called **gravity**. Gravity pulls every object toward every other object. Gravity is the force that pulls us toward Earth's surface, stopping us from floating away.

9

HOW FAR?

Distances in space are enormous. **Astronomers** measure the distance between two space objects by working out how long it would take for **light** to travel between them.

The speed of light

Light travels incredibly fast—more than 186,000 miles per second. It is the fastest thing in the Universe, so astronomers use its speed as a special measurement.

Astronomers can work out how long it takes for the starlight to reach Earth and therefore how far away the stars are.

MOON
Distance from Earth: 1.3 light seconds.

SUN
Distance from Earth: 8 light minutes.

Earth

8 minutes ago

Light years

It would take one year to travel from Earth to the Sun, 93 million miles away. It only takes eight minutes for the Sun's light to reach Earth. In one year, light travels 5,900,000,000,000 miles. This is called a **light year**.

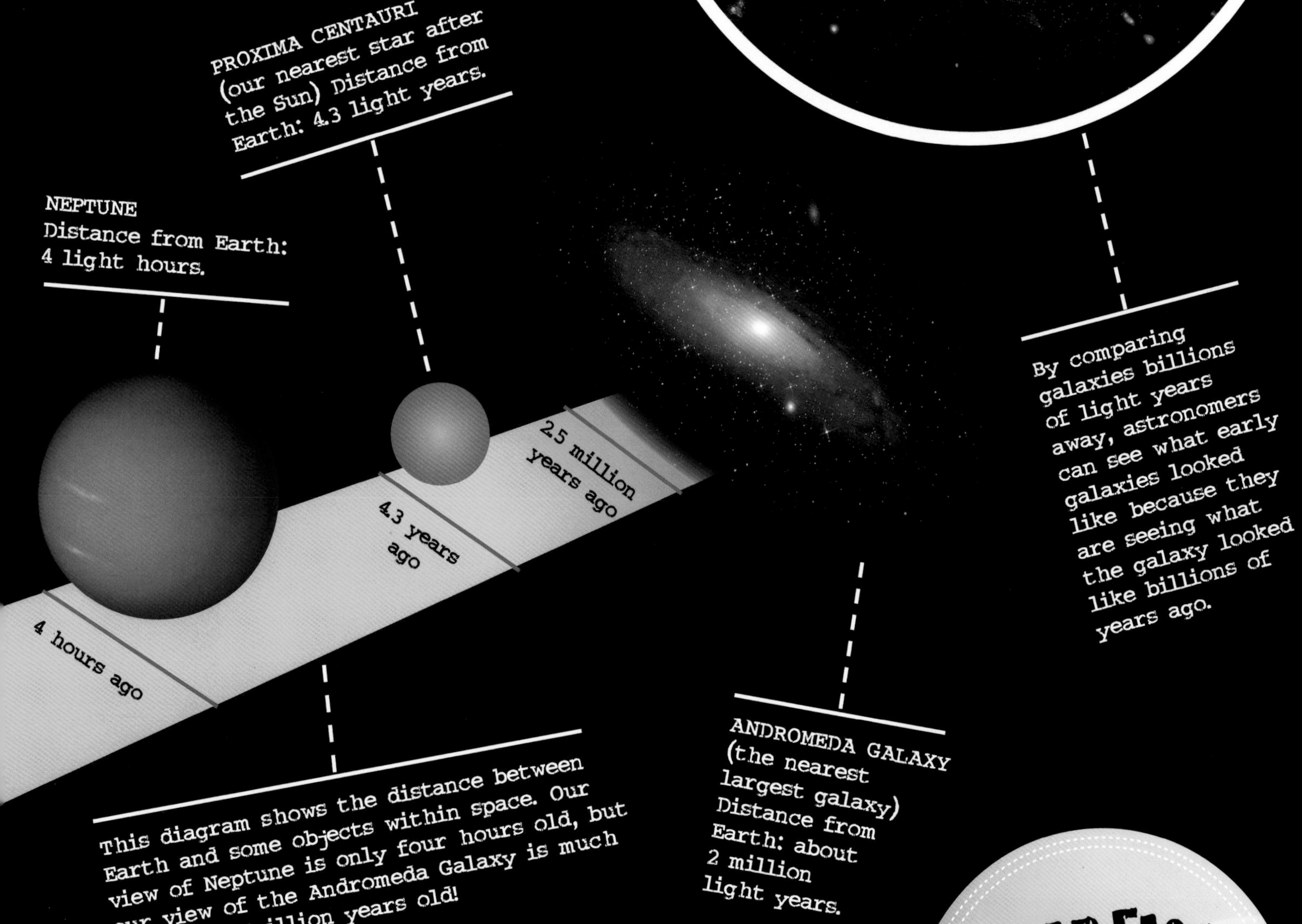

PROXIMA CENTAURI (our nearest star after the Sun) Distance from Earth: 4.3 light years.

NEPTUNE Distance from Earth: 4 light hours.

25 million years ago

4.3 years ago

4 hours ago

By comparing galaxies billions of light years away, astronomers can see what early galaxies looked like because they are seeing what the galaxy looked like billions of years ago.

This diagram shows the distance between Earth and some objects within space. Our view of Neptune is only four hours old, but our view of the Andromeda Galaxy is much older—two million years old!

ANDROMEDA GALAXY (the nearest largest galaxy) Distance from Earth: about 2 million light years.

Looking back in time

The farther away a space object such as a star is, the longer its light takes to reach Earth. The nearest star to Earth is the Sun. The second nearest star is more than four light years away. It takes four years for the star's light to reach Earth. This means that we are actually seeing what the star looked like four years ago!

STAR FACT

The farthest galaxy from Earth is 13.3 billion light years away. It was discovered in 2012 and is called MACS0647-JD.

FOCUS ON... ALBERT EINSTEIN

Physicist Albert Einstein (1879–1955) was one of the greatest scientific minds of all time. He is famous for the two theories of relativity, which helped us to better understand how the universe works.

Einstein won the 1921 Nobel Prize in Physics.

STAR FACT

Einstein learned to play the violin from the age of six.

Earth

Understanding relativity

Everything in the Universe is always moving, so the movement of an object can always be compared to the movement of another object. For example, if an asteroid passes Earth at a speed of 10,000 miles per hour. Then a spacecraft passes Earth at 15,000 miles per hour. The spacecraft is only traveling at a speed of 5,000 miles per hour relative to the asteroid.

Spacecraft travels at 15,000 miles per hour.

Testing theories

In 1916, Einstein published his theory of general relativity. This theory helped describe how gravity works. It said that the force of gravity could bend light. Therefore, the gravity of planets and stars made them act like giant **lenses**. Three years later, astrophysicist Arthur Eddington tested Einstein's theory and proved that it was correct.

Eddington took photographs of some stars during a solar eclipse. He compared these to photographs of the stars when the Sun was in another part of the sky. The stars seemed to have moved. This proved that the Sun was bending the light from the stars.

Asteroid travels at 10,000 miles per hour.

Relative speed of the spacecraft to the asteroid is only 5,000 miles per hour.

TIMELINE OF EINSTEIN'S LIFE

14 March 1879	Born in Ulm, Germany
1896	Studies math at Zurich Polytechnic
1915	General theory of relativity is complete
1919	Arthur Eddington proves the general theory of relativity is correct
1922	Receives the 1921 Nobel Prize in Physics
1955	Dies of a heart attack in the hospital

WHAT ARE GALAXIES?

A galaxy is a family of billions of stars that are held together by a special force called gravity. The galaxy that we live in is called the Milky Way. Galaxies can be all shapes and sizes and are spread out throughout space.

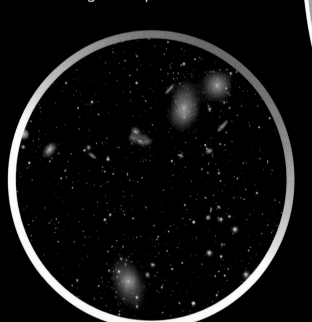

The Local Group of galaxies

The Virgo Cluster

The Virgo Cluster is a family of thousands of galaxies and measures 50 million light years across.

Grouped together

Large families of galaxies are called **clusters**. The Milky Way is part of a cluster of galaxies called the Local Group. Many clusters of galaxies are grouped into even bigger families called superclusters. A supercluster contains about 12 galaxy clusters and measures 100 million light years across. There are vast areas of empty space between each supercluster.

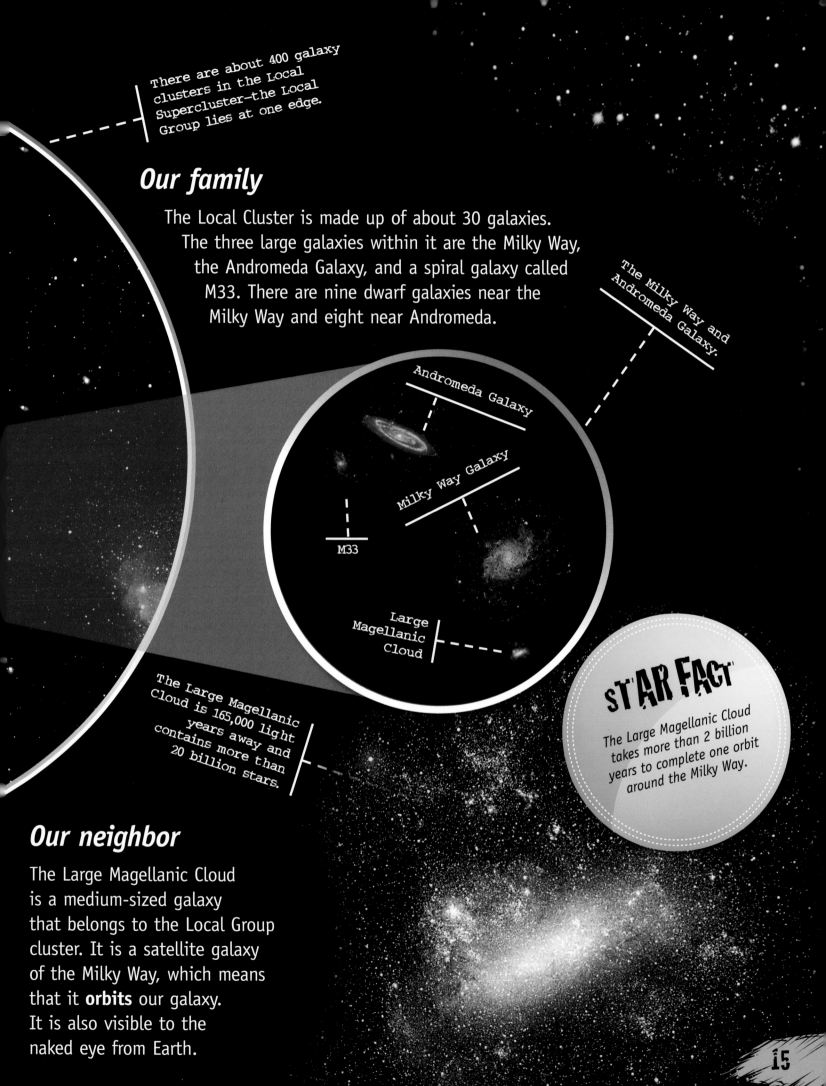

There are about 400 galaxy clusters in the Local Supercluster—the Local Group lies at one edge.

Our family

The Local Cluster is made up of about 30 galaxies. The three large galaxies within it are the Milky Way, the Andromeda Galaxy, and a spiral galaxy called M33. There are nine dwarf galaxies near the Milky Way and eight near Andromeda.

The Milky Way and Andromeda Galaxy.

Andromeda Galaxy

Milky Way Galaxy

M33

Large Magellanic Cloud

The Large Magellanic Cloud is 165,000 light years away and contains more than 20 billion stars.

STAR FACT

The Large Magellanic Cloud takes more than 2 billion years to complete one orbit around the Milky Way.

Our neighbor

The Large Magellanic Cloud is a medium-sized galaxy that belongs to the Local Group cluster. It is a satellite galaxy of the Milky Way, which means that it **orbits** our galaxy. It is also visible to the naked eye from Earth.

15

ALL SHAPES AND SIZES

There are billions of galaxies in the Universe, with different sizes, shapes, and masses. Some galaxies are faint and some give out great amounts of heat and light. The tiniest galaxies have a few million stars and the largest have several billion stars.

Egg shaped

Most galaxies are **elliptical**, shaped like an egg. Giant elliptical galaxies are rare and have ten times as many stars as the Milky Way. Most elliptical galaxies are dwarfs, with just a few million stars.

Leo 1 is a dwarf elliptical galaxy, only 750,000 light years from Earth. It is only 1,000 light years across.

The giant elliptical galaxy M87 is shaped like a sphere and has the same mass as 10 billion billion Suns.

STAR FACT

At least six galaxies are visible to the naked eye from Earth. They look like faint, misty patches of light.

This spiral galaxy looks similar to the Milky Way. Here, two spiral arms come from the center and contain young stars, glowing clouds of gas, and streaks of dust.

In a spin

Spiral galaxies are bigger than dwarf elliptical galaxies, with between one billion and one trillion stars. They usually have two arms spiraling out from a bulging center. As the galaxy turns, the arms seem to trail along.

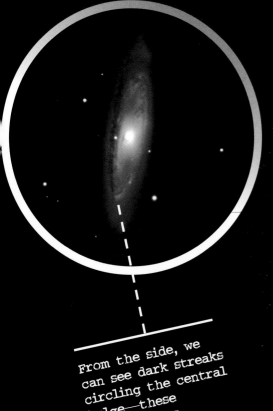

In this galaxy, the arms come from the central, elongated bar of stars, rather than a round center.

From the side, we can see dark streaks circling the central bulge—these are the arms.

Random arrangement

About one third of galaxies do not have a regular shape. They are called irregular galaxies and are usually smaller than spiral galaxies. Irregular galaxies contain a lot of gas. Sometimes they have become irregular due to the pull of a nearby galaxy's gravity.

This galaxy was spiral shaped until new stars were born.

17

CRASH! BANG!

Regions of large galaxy clusters can become very crowded. Galaxies can end up so close to one another that they collide. It takes millions of years for a galaxy to crash through another galaxy.

When galaxies collide

When two galaxies collide, the pull of gravity of each galaxy distorts the disks and spiral arms. This can cause long, spindly streamers of stars to fly off into space.

2

The galaxies move very close to each other, just before a collision. This has taken millions of years.

1

The galaxies in space.

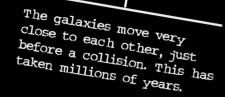

STAR FACT

If gas clouds crash into each other, they get squeezed, which rapidly forms new stars.

18

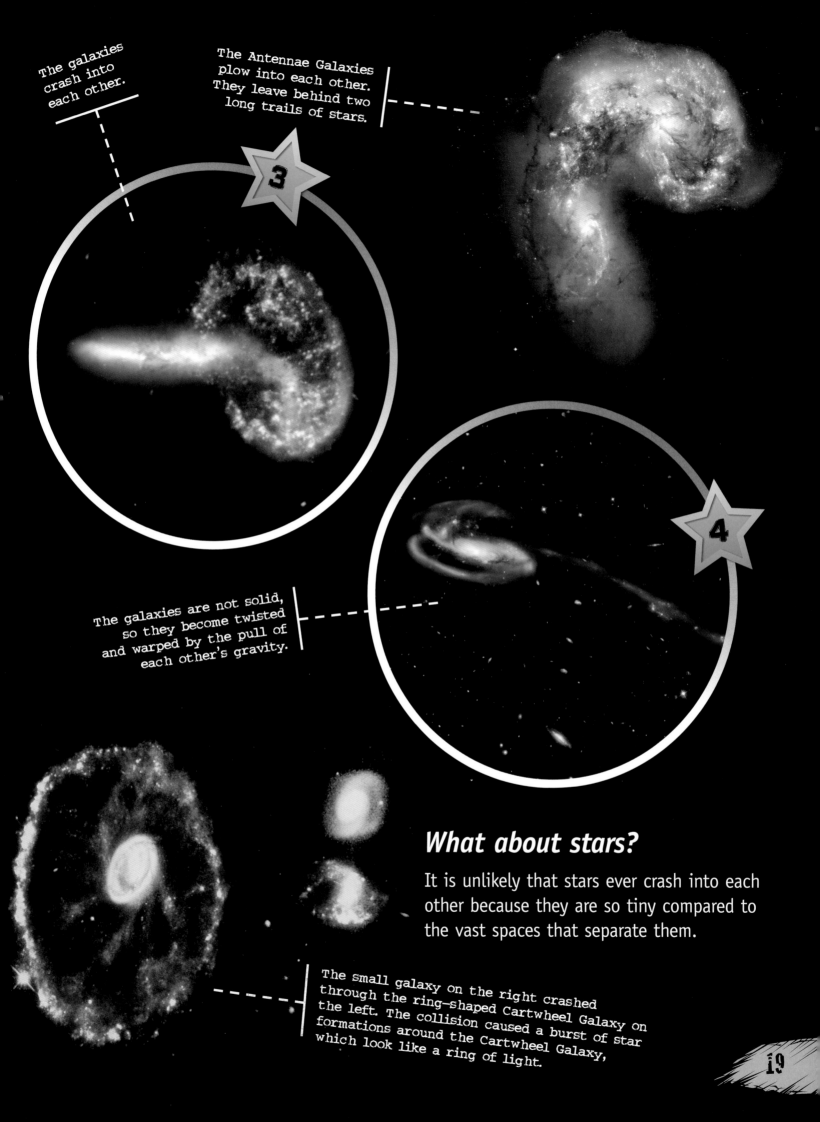

The galaxies crash into each other.

The Antennae Galaxies plow into each other. They leave behind two long trails of stars.

3

The galaxies are not solid, so they become twisted and warped by the pull of each other's gravity.

4

What about stars?

It is unlikely that stars ever crash into each other because they are so tiny compared to the vast spaces that separate them.

The small galaxy on the right crashed through the ring-shaped Cartwheel Galaxy on the left. The collision caused a burst of star formations around the Cartwheel Galaxy, which look like a ring of light.

BRIGHTEST GALAXIES

Light from galaxies usually comes from their stars. However, some galaxies have a different way of making light, which makes them very bright. They are called active galaxies.

Power center

Active galaxies produce huge amounts of energy from their centers, called quasars. They are the most energetic objects of the Universe, and are powered by **black holes**. Galaxies with quasars are much brighter than normal galaxies, so they can be detected much farther away.

The Very Large Array is a set of 27 radio telescopes in New Mexico. Each dish is 80 feet in diameter.

Radio galaxies

Astronomers discovered that some galaxies give off powerful **radio waves**. They are called radio galaxies and they often have two giant clouds of matter that lie outside the main galaxy. These clouds look like they have been blasted out of the galaxy. The radio waves mainly come from the clouds themselves.

On either side of this galaxy, there are jets coming from the center. These jets are giving out huge amounts of energy as radio waves.

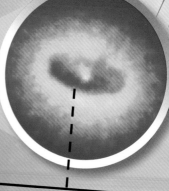

Near the center, a dark disk of gas swirls around the bright core. It measures 300 light years across.

A quasar looks like a bright star, but it is actually the center of a galaxy at least 100 times brighter than a normal galaxy. The galaxy surrounding the quasar is often too faint to be seen. Any extra light pours out of the quasar at its center.

There are two bright, narrow jets of matter and energy given off from either side of the quasar's core.

STAR FACT

The nearest quasar to Earth is more than one billion light years away.

THE MILKY WAY

The galaxy that we live in is called the Milky Way. It is a spiral galaxy, made up of 100 billion stars as well as clouds of dust and gas. The galaxy measures 150,000 light years across.

STAR FACT

It takes 220 million years for the Sun to complete one orbit through the Milky Way.

There is more to the Milky Way than astronomers can see. Dust covers the center of the galaxy, but astronomers have been able to map the galaxy by looking at the radio waves it gives off.

Making a galaxy

About 14 billion years ago, gravity pulled a giant, slowly-turning ball of gas inward until the ball collapsed to form a disk with a bulge in the middle. This formed our galaxy, the Milky Way. In the center of the Milky Way is a supermassive black hole.

The Milky Way is shaped like a thin disk with a central bulge and spiral arms. It contains billions of stars, which are packed much more closely together in the bulge than in the arms.

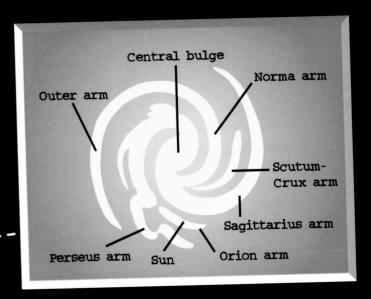

Central bulge

Norma arm

Outer arm

Scutum-Crux arm

Sagittarius arm

Perseus arm   Sun   Orion arm

The Milky Way looks like a faint band of light stretching across the night sky.

Spiraling stars

Astronomers have mapped the Milky Way and know that it is like any other spiral galaxy—a central bulge and curving arms with gas clouds and young stars. A sphere of stars called the halo sit around the galaxy. These stars have existed since before the galaxy's formation.

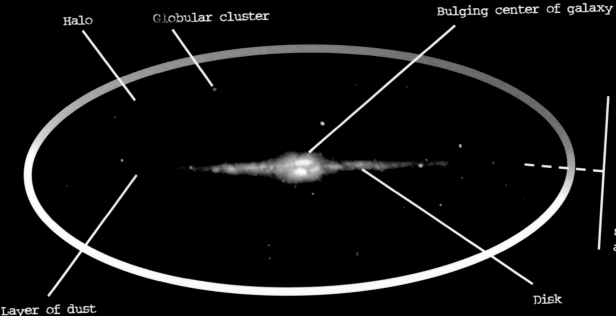

Halo   Globular cluster   Bulging center of galaxy

The halo surrounding the galaxy contains about 200 star clusters. Their position shows that the Milky Way was once shaped like a sphere.

Layer of dust   Disk

FOCUS ON...
EDWIN HUBBLE

One of the greatest astronomers of the 20th century, Edwin Hubble (1889–1953), proved there were galaxies beyond the Milky Way. The Hubble Space Telescope is named after him.

Hubble started work at the Mount Wilson Observatory, California, in 1919. He stayed there for the rest of his working life.

STAR FACT

Edwin Hubble had degrees in **astronomy** and law.

The speed of galaxies

After Hubble proved that there are more galaxies in space than our own, he then showed that these galaxies are moving away. When a galaxy is moving, the light it gives off changes. The light from a galaxy can be used to tell how far away it is. This change in light is known as the Doppler effect.

A galaxy that is coming nearer is said to have "blueshift."

A galaxy that is moving away is said to have "redshift."

Hubble's Law

Using the Doppler effect, Hubble measured the speeds of the newly discovered galaxies. He found that the farther away a galaxy is, the faster it is moving. The relationship between a galaxy's distance and its speed is called Hubble's law. It showed that the Universe is getting bigger.

Hubble's law shows that the Universe started very small and is expanding. This led to the idea of the Big Bang.

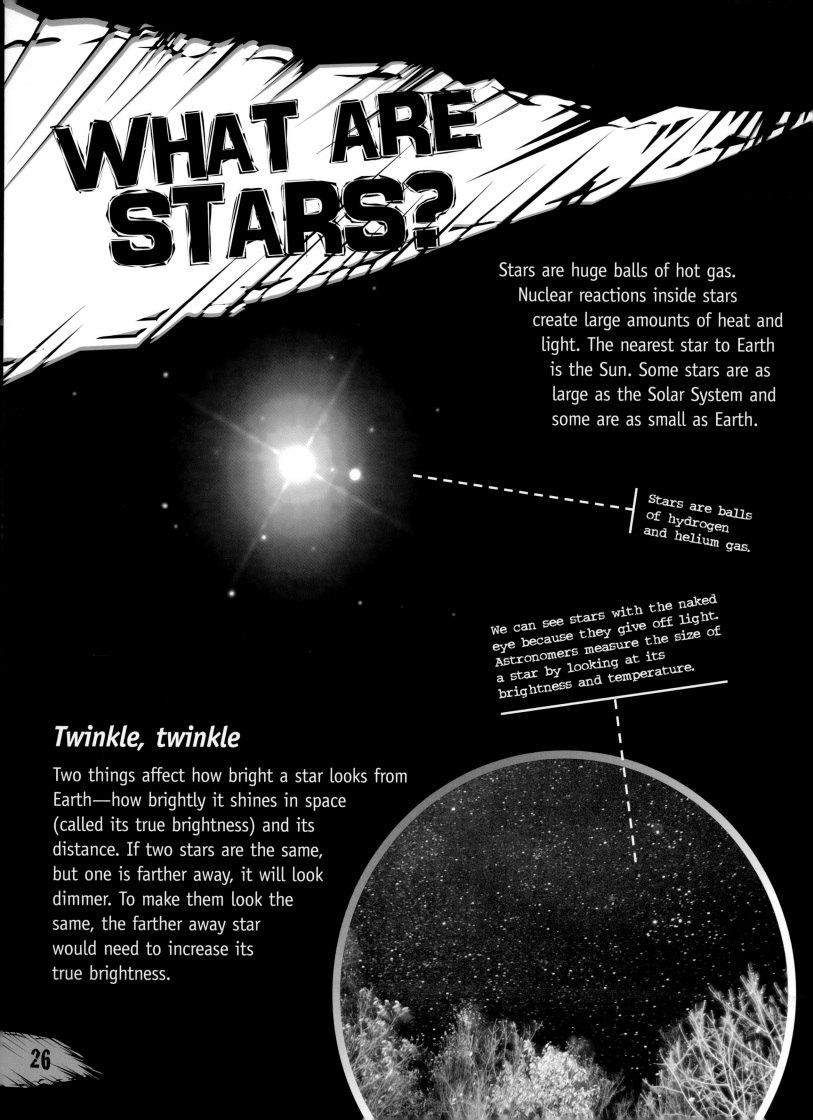

WHAT ARE STARS?

Stars are huge balls of hot gas. Nuclear reactions inside stars create large amounts of heat and light. The nearest star to Earth is the Sun. Some stars are as large as the Solar System and some are as small as Earth.

Stars are balls of hydrogen and helium gas.

We can see stars with the naked eye because they give off light. Astronomers measure the size of a star by looking at its brightness and temperature.

Twinkle, twinkle

Two things affect how bright a star looks from Earth—how brightly it shines in space (called its true brightness) and its distance. If two stars are the same, but one is farther away, it will look dimmer. To make them look the same, the farther away star would need to increase its true brightness.

Changing faces

Many stars can become brighter and dimmer in their lifetimes. They are called variable stars. Pulsating stars, a type of variable star, vary because the whole star expands and shrinks in a regular cycle. They help astronomers to measure distances to galaxies beyond the Milky Way.

As it goes through its cycle, the star's size, color, and temperature change.

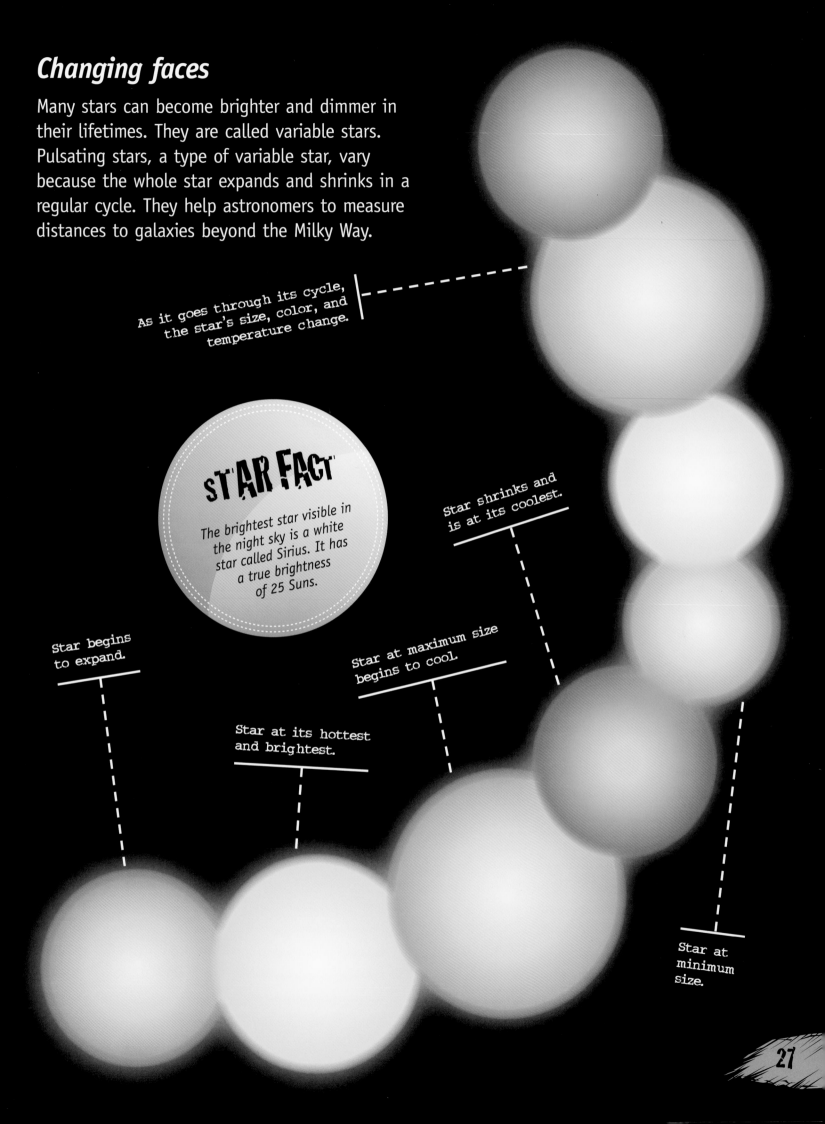

STAR FACT

The brightest star visible in the night sky is a white star called Sirius. It has a true brightness of 25 Suns.

Star shrinks and is at its coolest.

Star begins to expand.

Star at maximum size begins to cool.

Star at its hottest and brightest.

Star at minimum size.

27

THE SUN

The Sun is the star at the center of our Solar System. It provides Earth with vital heat and light needed for life on the planet. It is only 93 million miles from Earth, so astronomers are able to investigate the Sun in detail.

Earth

Hot, hot, hot

The Sun's surface is called the photosphere and its temperature is about 10,000 degrees Fahrenheit. The layer of **atmosphere** surrounding the Sun is called the chromosphere, at about 45,000 degrees Fahrenheit.

Deep inside the Sun, at the core, the temperature is a scorching 27 million degrees Fahrenheit.

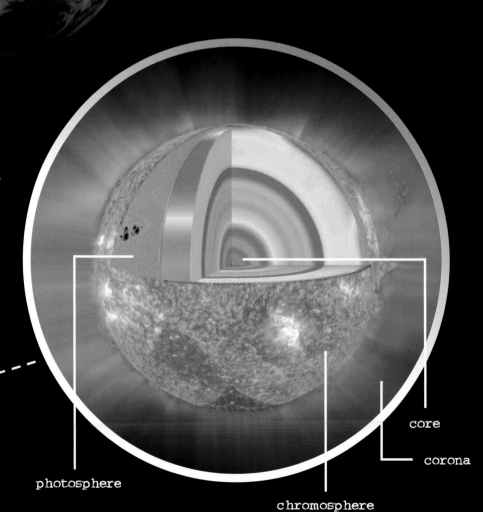

core

corona

photosphere

chromosphere

28

Astronomers use special telescopes and space observatories to carefully study the Sun so that they can monitor how its changes affect life on Earth.

Ring of fire

Stretching for millions of miles beyond the chromosphere is the corona. It is the hottest part of the Sun's surface, reaching temperatures of 2 million degrees Fahrenheit.

WARNING!

Never look straight at the Sun. It is so bright that it can damage your eyes.

Gases from the corona form solar winds, which blow into space at a speed of 500 miles per hour.

THE SUN'S FEATURES

The Sun's hot, glowing gases are always on the move. This creates a very active surface bubbling with explosions, flares, spikes, and spots.

The Sun's surface seems to bubble, giving it a mottled appearance called granulation.

STAR FACT

The Sun gets 4 million tons lighter every second because it releases so much energy!

Solar flares

Sudden, small explosions of energy from the Sun are called **solar** flares. They flare up in just a few minutes and take less than an hour to die away. They can reach temperatures of up to 18 million degrees Fahrenheit.

Solar flares blast particles out into space. Smaller spikes called spicules last only a few minutes, but are thousands of miles high.

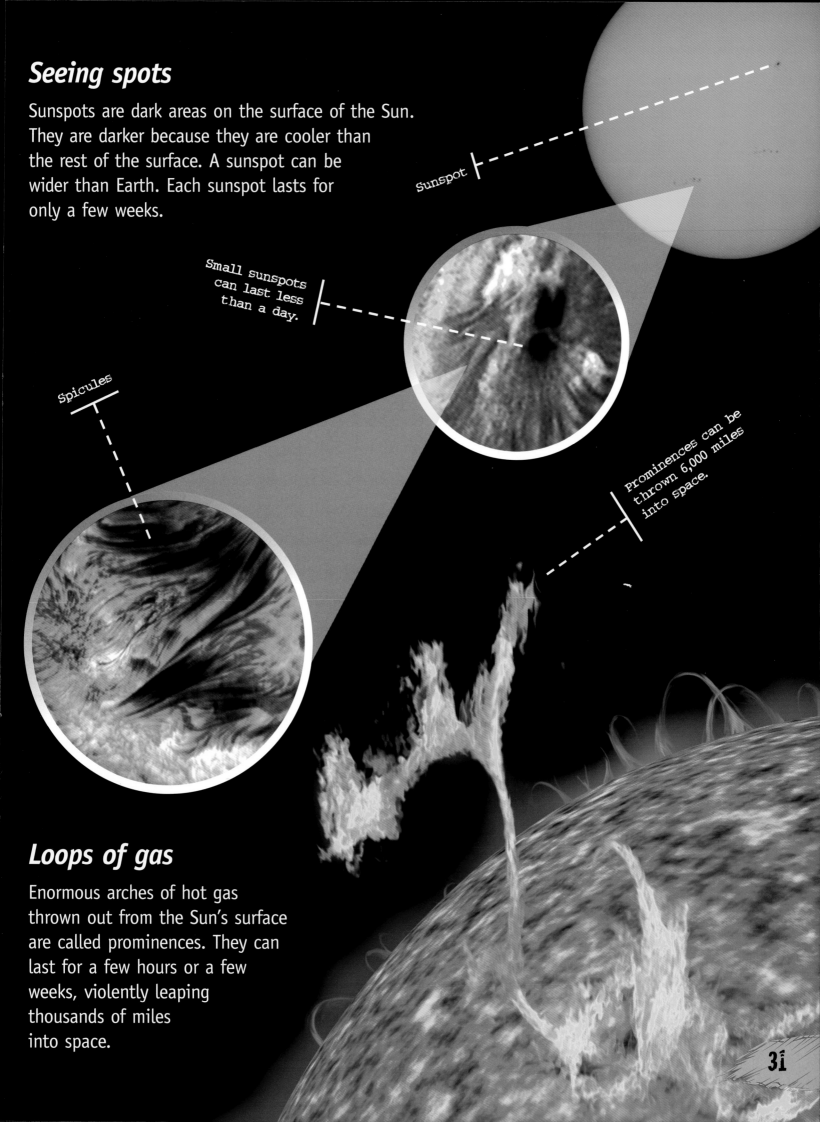

Seeing spots

Sunspots are dark areas on the surface of the Sun. They are darker because they are cooler than the rest of the surface. A sunspot can be wider than Earth. Each sunspot lasts for only a few weeks.

Sunspot

Small sunspots can last less than a day.

Spicules

Prominences can be thrown 6,000 miles into space.

Loops of gas

Enormous arches of hot gas thrown out from the Sun's surface are called prominences. They can last for a few hours or a few weeks, violently leaping thousands of miles into space.

31

SOLAR ECLIPSES

Every now and then, the Moon, Sun, and Earth line up in space. If the Moon is directly between the Sun and Earth, it stops sunlight from reaching a small area on Earth. This is called a solar eclipse.

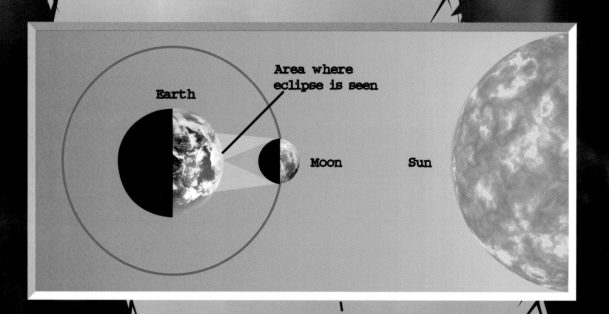

Area where eclipse is seen

Earth

Moon

Sun

When the Moon moves directly between the Sun and Earth, its shadow falls on Earth and there is a solar eclipse.

Sizing up

The Sun is 400 times larger than the Moon, but it is also 400 times farther away, so the Sun and Moon appear to be roughly the same size in the sky. There are between two and seven solar eclipses a year, but most of them are only partial, which means that the Sun isn't completely blocked by the Moon.

Each eclipse can be seen only from a small area on Earth, depending on how the Sun, Moon, and Earth line up.

WARNING!

Special glasses must be used to view eclipses of the Sun.

Total darkness

When the Moon completely covers the Sun, it is called a total eclipse. The sky turns dark and only a faint light from the Sun's corona can be seen. It looks like a white halo around the dark circle of the Moon. Tongues of gas leaping from the edge of the Sun may also be visible.

When the eclipse ends, a beam of sunlight bursts through and it looks like a diamond ring.

Partial eclipse   Total eclipse

STAR BIRTH

New stars form all the time. Stars that are born in huge, dark clouds of cold gas and dust are called **nebulae.** At first, the forming stars are hidden behind clouds of dust. Once stars are fully formed, their light or radiation breaks through the darkness and lights up the area around them.

Making a star

A star begins when gravity pulls clumps of gas together. The gas becomes a tight, hot ball. The ball gets hotter and hotter until nuclear reactions start to take place in its center, and a star is born.

These bright stars are only 50 million years old and can be seen with the naked eye in the constellation Taurus.

Clustering together

Stars always form together in clusters. A cluster can contain hundreds of stars or just a few. When astronomers see clusters in the sky, they know that they formed together. Over time, the stars in a cluster drift apart. All the stars throughout the Milky Way were once in clusters.

1

2

3

A cluster of young stars form in a dark cloud of dust and gas.

The first cluster of stars begins to spread out.

Hot gas from the new stars causes more stars to form in a nearby cloud.

A cloud of stars

The **constellation** Orion (the Hunter) forms a bright pattern in the sky. Below the three stars of Orion's Belt lies a misty patch of light. The glowing cloud of gas is called the Orion Nebula. Inside the Orion Nebula are four hot, young stars called the Trapezium. They are only 100,000 years old. Nearby, astronomers can see disk-shaped clouds of gas and dust. They believe that these disks may form new planets.

Orion constellation

Orion Nebula

Trapezium

New planets

STAR FACT

A ball of gas needs to reach 18 million degrees Fahrenheit before star birth can take place.

LARGE AND SMALL

Stars can be small or large, faint or bright, and different colors. Astronomers put stars into categories depending on their colors and sizes.

Size matters

Small stars are called dwarfs, large stars are giants, and the biggest are called supergiants. The Sun is a medium-sized star, twenty times larger than a white dwarf star. A supergiant is hundreds of times bigger than the Sun.

Red hot

A star's color shows its temperature as its surface gas heats up or cools down. Red stars are the coolest and blue stars are the hottest. The Sun is white but appears as yellow, so it's in between.

STAR FACT

The hottest known stars reach a super-sizzling temperature of 450,000 degrees Fahrenheit.

A red dwarf is half the width of the Sun, but is much dimmer. Its surface temperature is 6,000 degrees Fahrenheit.

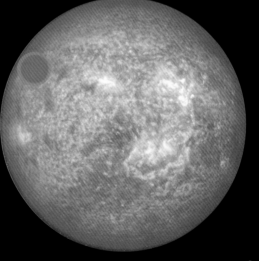

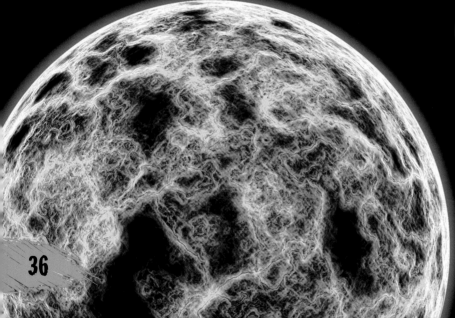

The Sun is larger than most stars in the Milky Way. Its surface temperature is 10,000 degrees Fahrenheit.

Gigantic stars

Some stars begin their lives much bigger than others. If large amounts of gas clump together as a single star, a young, very hot, blue giant star is produced. Red giant stars are enormous, but much older. They are cooler than blue giants because their heat has spread out over a larger area.

A white dwarf is an old, dying star. Its surface temperature is 45,000 degrees Fahrenheit.

A brown dwarf isn't a proper star because the ball of gas it formed from was so small. Its surface temperature is 1,300 degrees Fahrenheit.

Ten Suns would fit across the width of a blue giant. It is as bright as 800,000 Suns and has a surface temperature of 80,000 degrees Fahrenheit.

A red supergiant is as wide as 500 Suns and as bright as 40,000 Suns. Its surface temperature is 6,100 degrees Fahrenheit.

DEATH OF A STAR

Stars do not shine forever because eventually they run out of gas to make energy (heat and light). When this happens, the star's center collapses inward, making it hotter. The rest of the star expands outward, turning the star into a giant or supergiant.

DEATH OF A STAR WITH THE MASS OF 1 SUN

Ordinary star similar to the Sun.

Large blue star with 10 times the Sun's mass.

DEATH OF A STAR WITH THE MASS OF 10 SUNS

Rings of gas were blown off this star before it exploded.

Blue supergiant.

Giant to dwarf

Smaller stars, such as the Sun, use up their gas more slowly than larger stars, so they last longer. They can shine for 10 billion years with hardly any change. At the end of their lives, they swell up into red giants.

Star becomes
a red giant

White dwarf gradually
cools, becoming dimmer
and redder.

Red giant blows
off shells of
gas—the collapsed
core of the star
is revealed at
the center as a
white dwarf.

A supernova explosion—
the outer layers of the
star are blasted off.

STAR FACT

New stars form
using the gas of old
stars. This means the
Universe is recycled!

Super blast

Heavy stars end their lives in a
massive explosion called a
supernova. The explosion blasts a
huge cloud of gas and dust into
space, leaving behind a tiny, hot
neutron star, or a black hole.

Blue supergiant
expands and
gets redder as
it gets older.

Star becomes a
red supergiant.

Collapsed core
becomes a neutron
star. If the star
has 20 times the
mass of the Sun or
more, it will form
a black hole.

39

BLACK HOLES

A star born with the mass of 20 suns will die as a black hole. The gravity inside a black hole is so strong that it sucks everything, including light, inside. Astronomers believe that black holes are likely to be at the center of every galaxy, including the Milky Way.

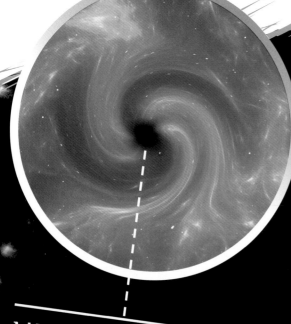

A black hole looks black because no light can escape from it.

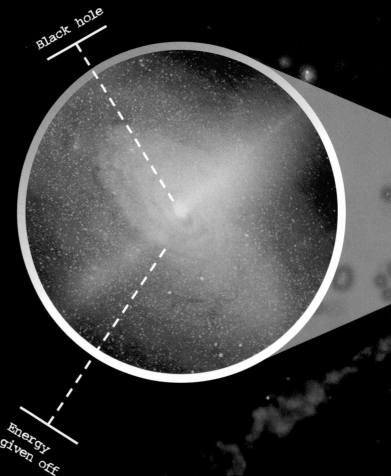

Black hole

Energy given off

Power center

When stars and gas are pulled into a black hole, a swirling disk of hot gas forms and enormous amounts of energy are given off. In the biggest galaxies, the black hole has grown to a few billion times the mass of the Sun.

This galaxy may be 100,000 light years across, but the swirling gases surrounding the black hole cover an area only a few hundred light years across.

Galaxy

Twin stars

A black hole on its own in space would be difficult to find. Sometimes if a black hole is very close to another star, they begin to work together. The black hole pulls streams of gas from the other star, which spiral around before being sucked into the black hole.

The black hole's gravity pulls gases from the giant blue star toward it.

WHAT IS THE SOLAR SYSTEM?

The Sun is surrounded by a family of eight planets and their moons, including Earth. The Sun's gravity pulls the planets, as well as thousands of asteroids and icy comets, toward it and keeps them orbiting around. This is the Solar System.

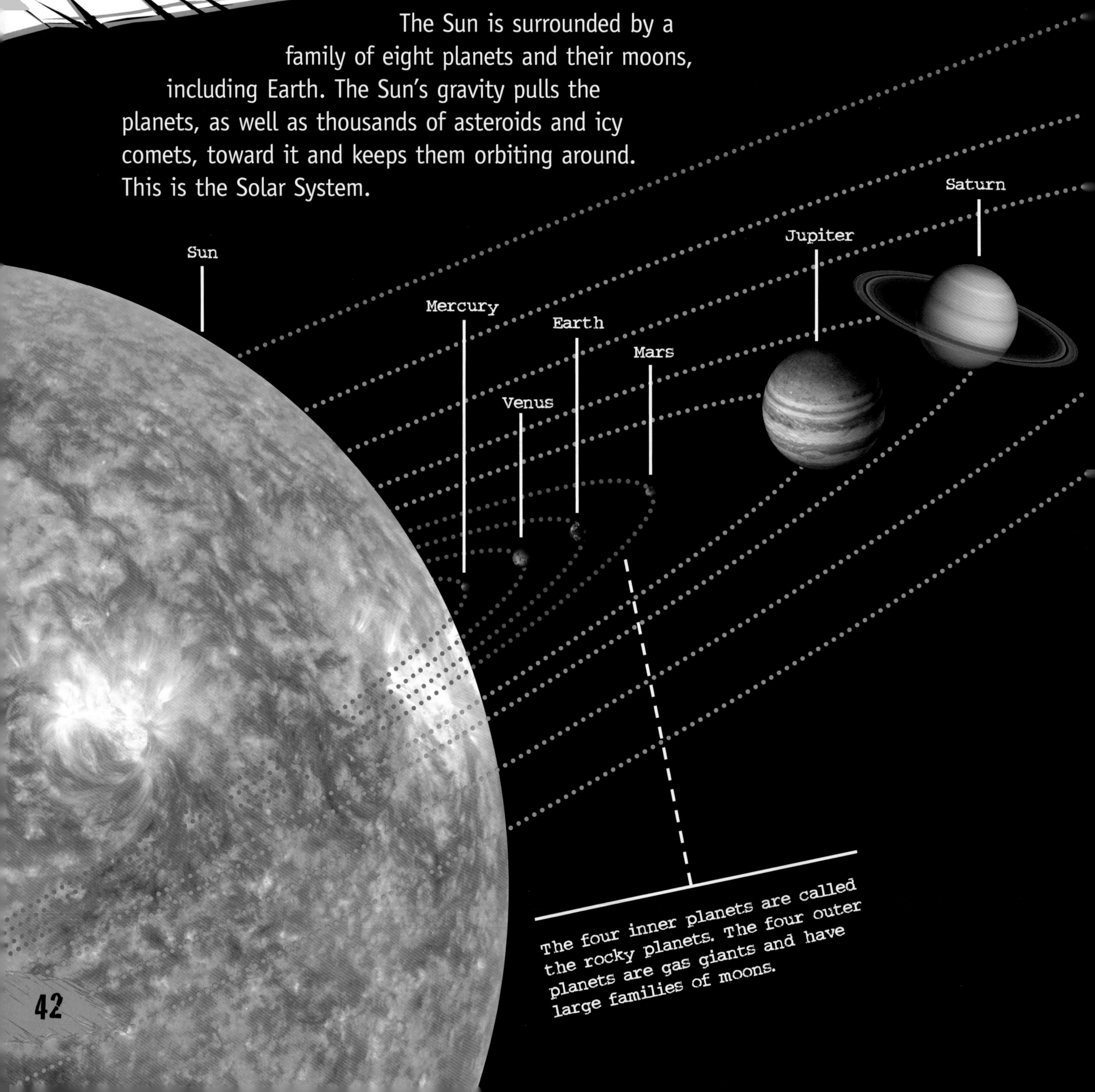

Sun

Mercury

Venus

Earth

Mars

Jupiter

Saturn

The four inner planets are called the rocky planets. The four outer planets are gas giants and have large families of moons.

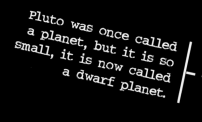

Pluto was once called a planet, but it is so small, it is now called a dwarf planet.

Neptune

Uranus

The family circle

The planets in the Solar System vary greatly in size and behavior. Mercury is closest to the Sun and is very small and hot. Earth, Mars, and Venus are cooler, but rocky. Jupiter, Saturn, Uranus, and Neptune are very large, cold, and made of gas.

The rest of the family

As well as the main planets, millions of other space objects are part of the Solar System. Thousands of minor rocky planets called asteroids orbit the Sun, including five known dwarf planets. Icy rocks called comets come from the outer Solar System.

Asteroids can mainly be found in the Asteroid Belt between Mars and Jupiter.

THE SOLAR SYSTEM BEGINS

The Solar System, including the Sun and planets, formed 4.6 billion years ago from the same slowly rotating cloud of gas and dust in space.

2

1

In a swirling cloud of gas and dust, gravity pulled material together to form a disk. The central part of the disk formed the young Sun.

Small clumps of material came together. Some of these collided with each other, smashing apart. Others merged together to make larger clumps. These large clumps were bombarded with leftover space rocks and dust.

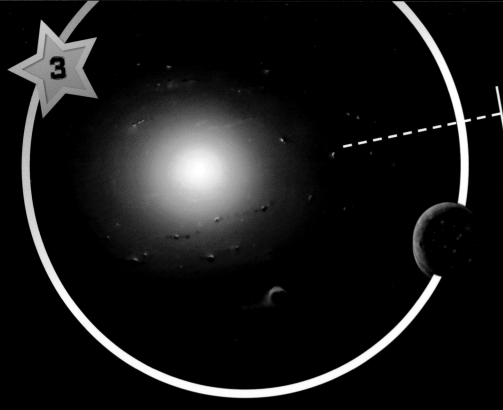

Over millions of years, these larger clumps of material became the planets. The smaller chunks became asteroids and comets.

Continuous change

The Solar System has changed a lot since it was formed and continues to do so. The planets have formed their own moons.

Making moons

Moons are balls of rock that orbit the planets, held in place by the planet's gravity. They can range from just a few miles across to more than 3,000 miles across. Some moons were simply lumps of rocks passing by a planet when they were captured by the planet.

Every planet in the Solar System has a moon, apart from Mercury and Venus.

ST'AR FACT

There are more than 182 known moons in the Solar System.

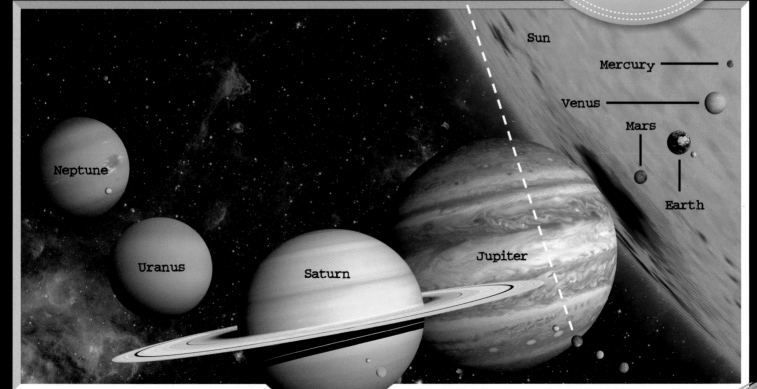

Sun

Mercury

Venus

Mars

Earth

Neptune

Uranus

Saturn

Jupiter

CIRCLING THE SUN

The Sun's gravity pulls everything in the Solar System toward it. The planets are speeding around, so they do not crash into the Sun. Instead they travel around the Sun, in orbits.

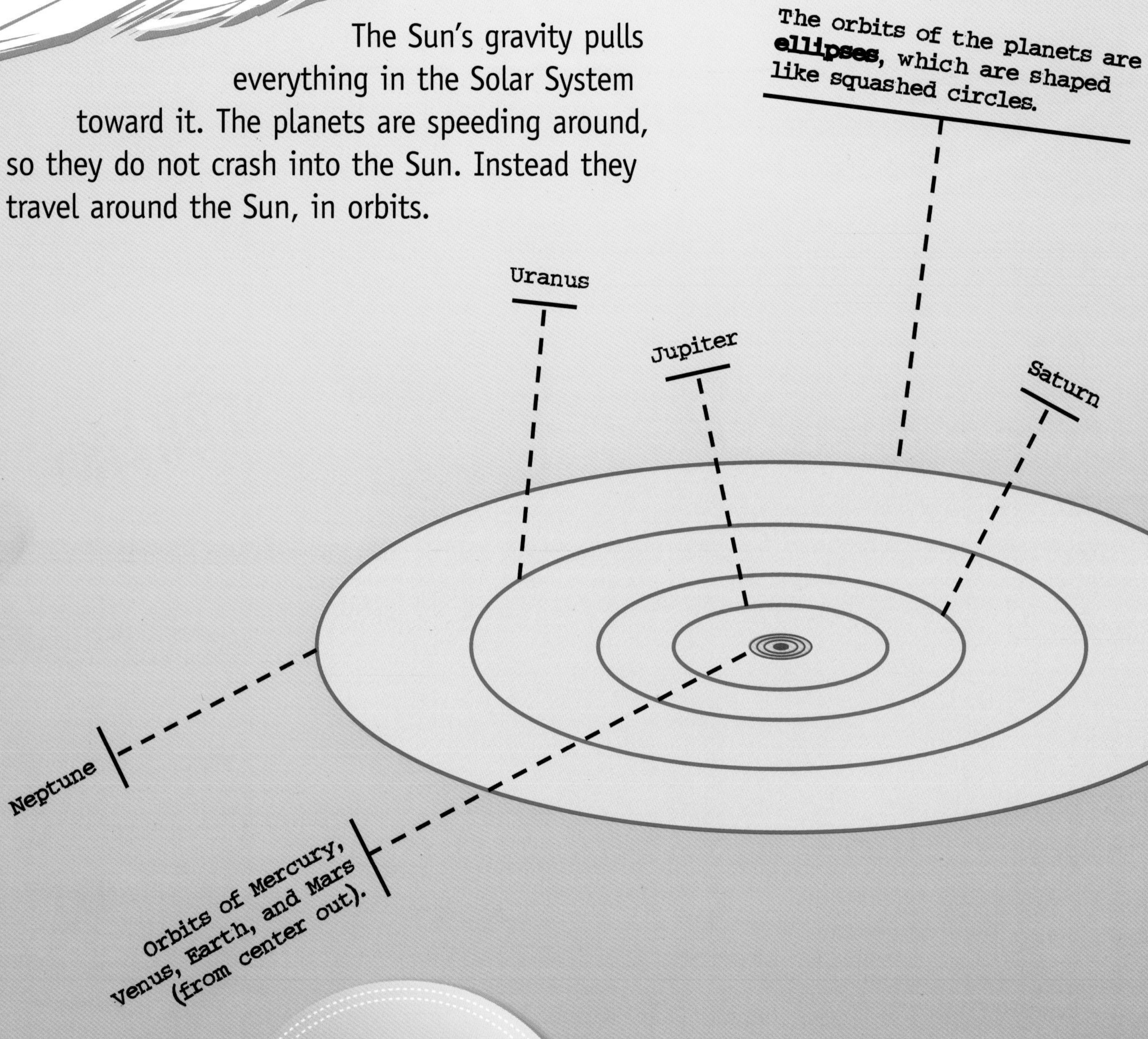

The orbits of the planets are **ellipses**, which are shaped like squashed circles.

Uranus

Jupiter

Saturn

Neptune

Orbits of Mercury, Venus, Earth, and Mars (from center out).

STAR FACT

Stars in the Milky Way can take up to 200 million years to orbit the center of the galaxy.

Neptune

Mercury

Great distances

The planets all travel at certain distances from the Sun, in fixed orbits. They travel slower the farther away they are from the Sun. Neptune is farthest from the Sun and Mercury is the closest.

Neptune takes more than 800 years longer to orbit the Sun than Mercury.

From close to far

The planets' orbits are elliptical, which means that sometimes they are quite close to the Sun and sometimes much farther away. Mercury gets closest to the Sun at only 28.5 million miles away. Neptune's farthest distance from the Sun is 2.8 billion miles.

FACT FILE

Planet	Time taken for one orbit around the Sun (Earth days/years)
Mercury	87.97 days
Venus	224.70 days
Earth	365.26 days
Mars	686.98 days
Jupiter	11.86 years
Saturn	29.46 years
Uranus	84.01 years
Neptune	164.8 years

Mercury gets very close to the Sun twice during its orbit. It is the fastest planet in the Solar System, reaching speeds of up to 30 miles per second.

FOCUS ON... JOHANNES KEPLER

German astronomer Johannes Kepler (1571–1630) discovered that the planets orbit around the Sun. He also worked out that the orbits are oval shaped. He developed three laws to explain his ideas.

Johannes Kepler also believed that the planets made music as they orbited the Sun.

Kepler also noted that the Sun is not in the center of the orbits. Instead, it is at one end, or "focus."

The first law

Before Kepler's discoveries, it was thought that the planets moved in circles. Kepler's first law states that the orbits of the planets are elliptical, or oval shaped.

Elliptical orbit

Planets move slower when they are farther away from the Sun.

The second law

Kepler's second law is about the speed at which the planets travel. He worked out that the speed of a planet depends on how close it is to the Sun. A planet moves faster when it travels nearer to the Sun. It is therefore slowest when it is farther away from the Sun.

Sun at one "focus"

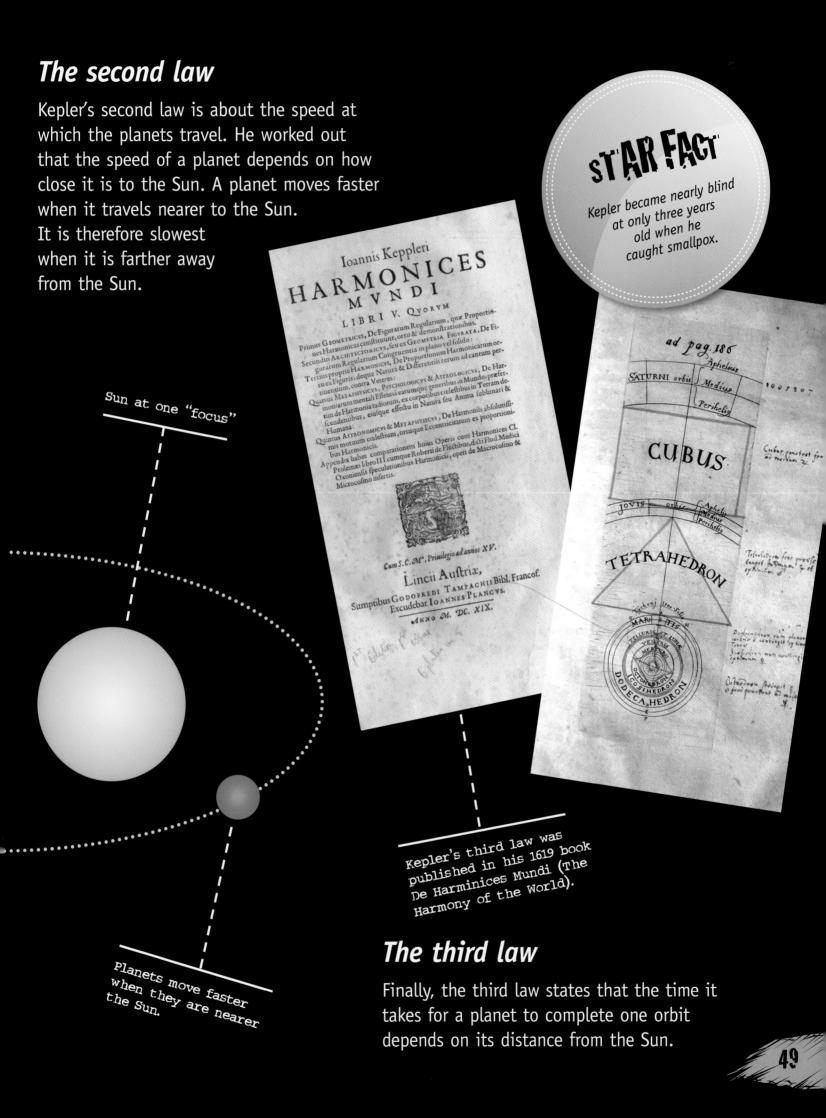

Planets move faster when they are nearer the Sun.

Kepler's third law was published in his 1619 book De Harminices Mundi (The Harmony of the World).

The third law

Finally, the third law states that the time it takes for a planet to complete one orbit depends on its distance from the Sun.

FOCUS ON... ISAAC NEWTON

Isaac Newton (1642–1727) is thought to be the greatest scientist and mathematician of all time. He discovered the force of gravity and the three laws of motion. He even explained why a rainbow is made up of many different colors.

The force of gravity

Newton came up with his ideas about gravity after seeing an apple fall from a tree when he was sitting in his yard. He went on to discover that the same force that gives humans weight and stops us from floating away, gravity, also makes the planets orbit the Sun, and the moons circle the planets.

In the late 1660s, Newton became a Professor at Cambridge University, England.

Newton discovered that if two space objects are light in weight, the pull of gravity between them is weak. If the objects are heavier, then the pull of gravity becomes stronger.

Using Newton's theories today

To work out the pull of gravity between two space objects, Newton came up with a calculation. This allowed astronomers to estimate the movement of stars and planets. Since then, astronomers have discovered unknown stars and planets, including Neptune.

Newton's ideas transformed our understanding of the Universe.

The Laws of Motion

Newton's laws of motion show how objects are able to move and stop, and how these motions can change when different forces are applied.

Law 1
Without an outside force, an object will travel in a straight line at the same speed without stopping, or it will never move in the first place.

Law 2
The more force that is applied to an object, the more acceleration it has. If the same force is applied to two objects, and one object is heavier than the other, the heavier object will accelerate more slowly.

Law 3
For every force, there is an opposite force. If you sit on a chair, you are pushing downward on the seat. The chair also has a force and pushes upward. If it didn't, the chair would collapse.

STAR FACT

Newton's discoveries helped make the first flights to the Moon possible.

51

THE ROCKY PLANETS

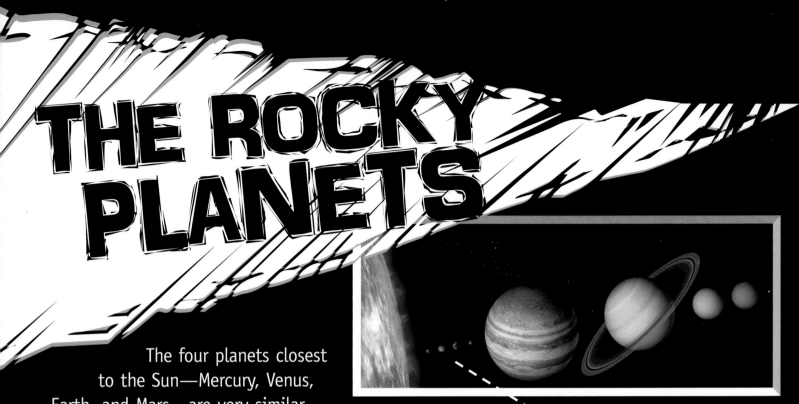

The four planets closest to the Sun—Mercury, Venus, Earth, and Mars—are very similar. When they formed 4.6 billion years ago, these planets kept crashing into chunks of rock and ice, giving them **cratered** surfaces. To begin with, the planets were very hot. They are made of rock and iron.

The inner planets are made mainly of rock and have solid surfaces.

Cores and crusts

Each rocky planet, as well as the Moon, has a heavy **core** of rock or iron. The cores of Mercury, Mars, and the Moon have cooled and hardened, but the cores of Venus and Earth are still liquid. On the outside of the planets, the solid crust is made of strong rock a few miles thick.

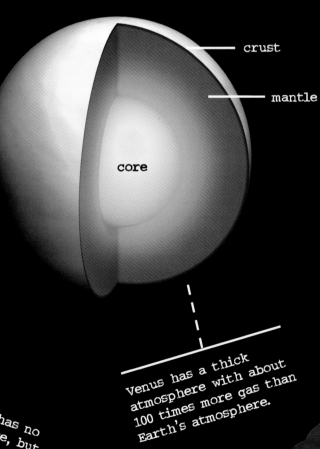

crust

mantle

core

Venus has a thick atmosphere with about 100 times more gas than Earth's atmosphere.

crust

mantle

core

Mercury has no atmosphere, but there is a small amount of gas around it.

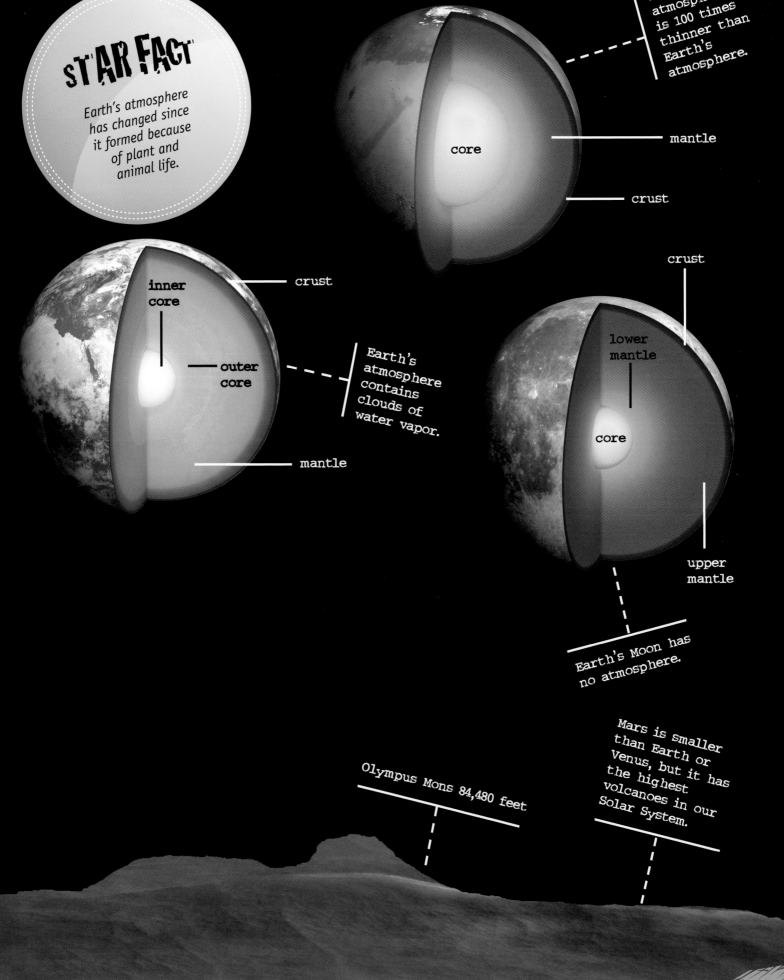

Earth's atmosphere has changed since it formed because of plant and animal life.

...atmosphere is 100 times thinner than Earth's atmosphere.

mantle

crust

core

crust

inner core

outer core

mantle

Earth's atmosphere contains clouds of water vapor.

lower mantle

core

crust

upper mantle

Earth's Moon has no atmosphere.

Olympus Mons 84,480 feet

Mars is smaller than Earth or Venus, but it has the highest volcanoes in our Solar System.

MERCURY, THE SMALLEST PLANET

Mercury is the smallest planet in our Solar System. It is also the closest planet to the Sun. Mercury takes 59 Earth days to spin around once. This means that it only spins 1.5 times for each orbit of the Sun. This makes a day on Mercury very long!

Cratered surface

Mercury is one of the most cratered objects in our Solar System. Its surface has been shaped by volcanic eruptions and by being crashed into by rocks. Mercury also looks wrinkled because the planet shrank a little when its surface cooled down.

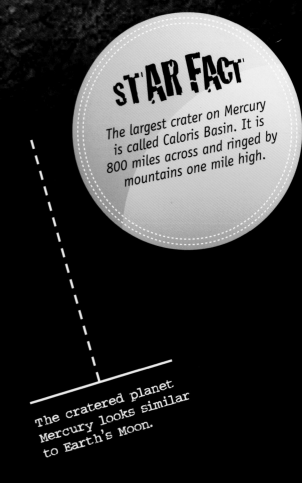

STAR FACT

The largest crater on Mercury is called Caloris Basin. It is 800 miles across and ringed by mountains one mile high.

The cratered planet Mercury looks similar to Earth's Moon.

Baking planet

Mercury has no atmosphere because it is very hot. It also has weak gravity, so cannot hold onto much gas. Mercury's daytime temperatures are about 750 degrees Fahrenheit, which is hot enough to melt lead. At nighttime, the temperatures plummet to -290 degrees Fahrenheit.

Mercury spins around so slowly that the side facing away from the Sun turns very cold, especially as there is no atmosphere to trap the heat.

How are craters formed?

Objects such as comets and rocks crash into the planet or moon's surface, creating **craters**. When a rock hits the surface, the impact causes the rock to explode. Small pieces of shattered rock are thrown upward from the crater. The pieces of rock fall back down. Some of them fill the crater and some start to form the circular wall around the crater.

Mercury has many large craters, some with raised centers. These are called peak rings.

VENUS, THE HOTHOUSE

Venus is very bright because its clouds reflect sunlight, making it the easiest planet to spot in the sky. It takes 243 Earth days for Venus to spin around once, giving it the longest day of any planet.

Morning and evening star

For a few weeks every seven months, Venus is the brightest object in the evening sky to the west. At that time, it is called the "evening star." About three-and-a-half months later, Venus appears in the eastern sky before sunrise. It is then called the "morning star."

Venus' surface is hidden from view by the thick, swirling clouds of its atmosphere.

Venus reflects almost 70% of the light it receives from the Sun.

Trapped heat

Venus has a hot, dry atmosphere. It is mainly made of carbon dioxide gas, which traps the Sun's heat. This makes the surface temperature of Venus extremely high, at 900 degrees Fahrenheit.

Floods of lava

Venus has hundreds of thousands of volcanoes. Most of them are about 2 miles across and 300 feet high. All over Venus, lava bursts through the thin, rocky crust, flowing in every direction.

As well as shield volcanoes, Venus also has extinct "dome volcanoes," which can only be found here. They formed after single eruptions of very sticky lava.

EARTH, THE LIVING PLANET

As far as scientists know, Earth is the only planet in the Universe where there is life. Earth is special because it has certain features, which all work together, to make it a habitable planet.

Earth has huge supplies of water, forming rivers, oceans, and seas. This water is vital, allowing living things to grow.

STAR FACT

About 30 miles above Earth's surface, a special kind of oxygen called ozone blocks harmful rays from the Sun.

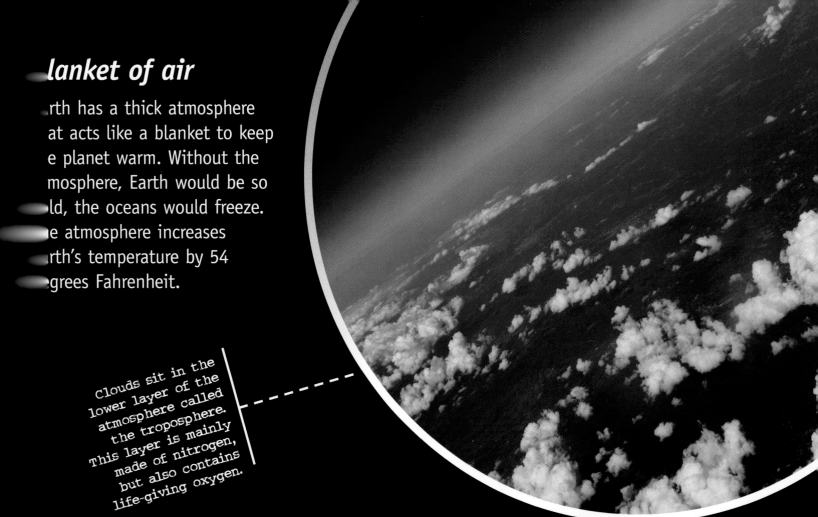

lanket of air

...rth has a thick atmosphere ...at acts like a blanket to keep ...e planet warm. Without the ...mosphere, Earth would be so ...ld, the oceans would freeze. ...e atmosphere increases ...rth's temperature by 54 ...grees Fahrenheit.

Clouds sit in the lower layer of the atmosphere called the troposphere. This layer is mainly made of nitrogen, but also contains life-giving oxygen.

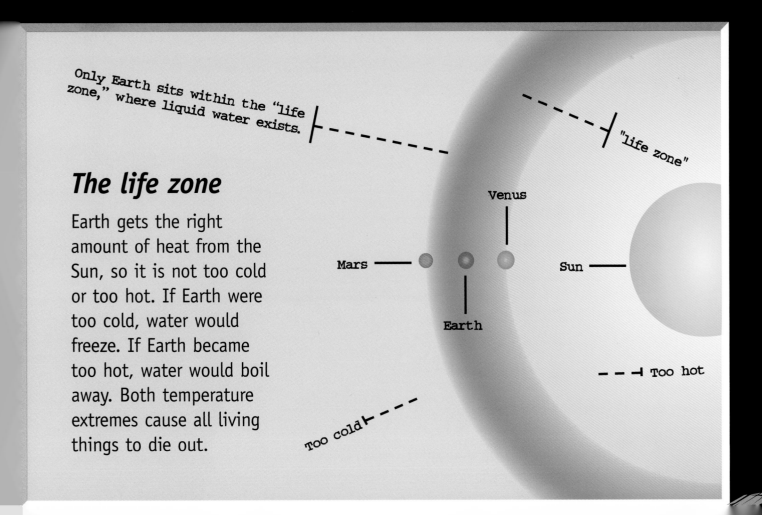

Only Earth sits within the "life zone," where liquid water exists.

"life zone"

Venus

Mars

Earth

Sun

Too hot

Too cold

The life zone

Earth gets the right amount of heat from the Sun, so it is not too cold or too hot. If Earth were too cold, water would freeze. If Earth became too hot, water would boil away. Both temperature extremes cause all living things to die out.

INSIDE EARTH

Earth is a large ball of rock, made of many layers. The outside layer is covered in land and water. This layer is where we live. Deep inside Earth, there is boiling hot liquid that is always moving. Sometimes we see this liquid when it erupts from volcanoes.

The inner core is made of super hot, solid iron metal.

The outer core is liquid iron metal

The thick layer of rock under the crust is the mantle.

The outer layer is called the crust.

Magma rises up through the crust from the mantle, and erupts out of volcanoes.

The moving mantle

Although the **mantle** is made of solid rock, it moves! The rock is extremely hot, creating huge amounts of **pressure**, so it flows, like very slow-moving molasses. The mantle is about 1,800 miles thick.

At the center

The incredibly hot temperature of the solid inner core creates so much pressure that the metal cannot melt. It measures 1,500 miles in diameter. The liquid outer core surrounds the inner core. As this liquid iron moves, it creates Earth's **magnetic** field.

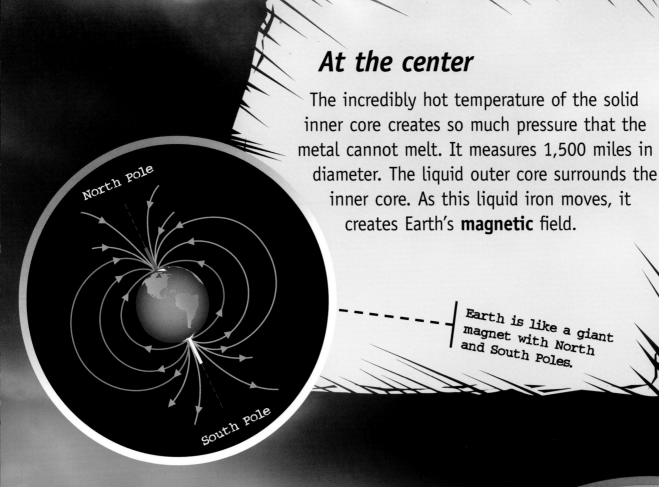

North Pole

South Pole

Earth is like a giant magnet with North and South Poles.

The crusty surface

The crust is the rocky surface where we live. It is split into pieces called plates. These plates float and slowly move on top of the mantle. The continents and ocean floors lie on top of these plates. Many of Earth's features, such as mountains and valleys, form when these plates move.

When Earth's plates crash together, huge ridges or mountain ranges form.

STAR FACT

Earth's plates move by a few inches every year.

SPINNING EARTH

Earth is constantly on the move. We measure a year by one of Earth's orbits around the Sun. We measure a day by one of Earth's rotations. Earth is also tilted, which gives us seasons.

A year on Earth

Earth takes about 365 days to complete one orbit of the Sun. This is Earth's year and is the time it takes for the Sun to be in exactly the same height in the sky.

STAR FACT

The Sun is at its highest point in the sky at solar noon each day.

Earth's orbit around the Sun is 584 million miles in length and actually takes 365.242 days.

Day and night

As Earth orbits the Sun, it also turns. It completes one turn every 24 hours, giving us a day. When one side of Earth is facing the Sun, it is daytime. At the same time on the other side of Earth, it is nighttime. As Earth slowly spins, the area that was facing the Sun turns away and it turns to nighttime. Night and day follow each other because Earth is always spinning.

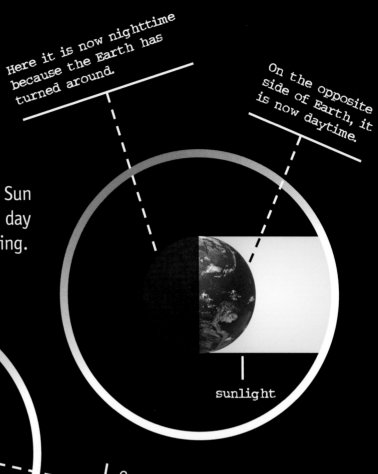

Here it is now nighttime because the Earth has turned around.

On the opposite side of Earth, it is now daytime.

sunlight

Here it is daytimee.

On the opposite side of Earth, it is nighttime.

sunlight

From summer to winter

As Earth orbits the Sun, it is tilted. This tilt causes the Sun's rays to hit some parts of the Earth more directly than others, giving these areas spring and summer. The areas of Earth that are farthest away from the Sun have fall and winter. The middle of Earth, called the **Equator**, stays almost the same.

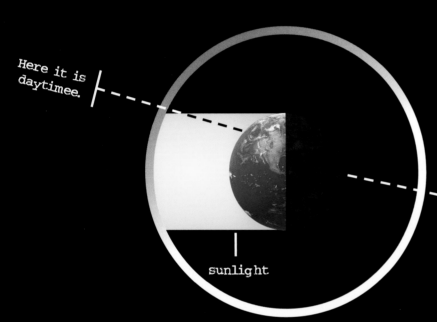

Spring is between winter and summer.

When the North Pole tilts away from the Sun, it is winter in the north and summer in the south.

When the North Pole tilts toward the Sun, it is summer in the north and winter in the south.

Fall is between summer and winter.

THE MOON

Astronauts have visited the Moon, Earth's closest neighbor in space. It is a ball of rock, covered in craters and dust. There is no life on the Moon because there is no liquid water and its gravity is so weak that all gases escape into space.

A night on the Moon lasts for 14 Earth days. The temperature drops to a chilly -300 degrees Fahrenheit.

STAR FACT

The largest crater on the Moon is Bailly. It is 183 miles across. That is the same length as 2,684 football fields!

Bashed to bits

Over time, the Moon has been bombarded by **meteors**, creating impact craters. Some craters are very large. Tycho is 53 miles across and Copernicus is 58 miles wide. When a meteorite hits, it breaks down into tiny pieces, covering the Moon in dust, especially around the impact zones.

The Eratosthenes crater is 36 miles across. The wall of dust around it that was created on impact has started to break down.

Crash into life

After Earth formed, it may have been hit by another rocky planet, causing rocks from both objects to fly into space. These rocks may have clumped together to create the Moon 4.6 billion years ago.

1 A planet as large as Mars collided with Earth.

2 Rocks from both planets were thrown into space.

3 These rocks joined together to form the Moon.

4 The new Moon began to orbit Earth.

Seas and mountains

Mountains were one of the first features that formed on the Moon, so they are now very old. Over time, large asteroids punched through the Moon's surface, causing lava to pour out. These areas were called "seas." The Moon has cooled down since then, so it is almost solid and there are no erupting volcanoes left.

Craters, seas (dark areas), and bright mountainous areas can be seen on the Moon by using only a pair of binoculars.

65

CHANGING SHAPES OF THE MOON

The Moon takes four weeks to complete one orbit of Earth. From night to night, the Moon's position in the sky changes as it moves from constellation to constellation. During this time, the shape of the Moon in the sky also seems to change. These are known as the phases of the Moon.

The phases

The Moon shines because it reflects sunlight. Only the side of the Moon that is facing the Sun can be seen. During a "Full Moon," we can see the whole Moon, but during a "New Moon" we cannot see the Moon at all.

The time between two New Moons is 29 days, 12 hours, and 44 minutes. When more of the Moon is becoming visible, it is "waxing" and when it starts to disappear again, the Moon is "waning."

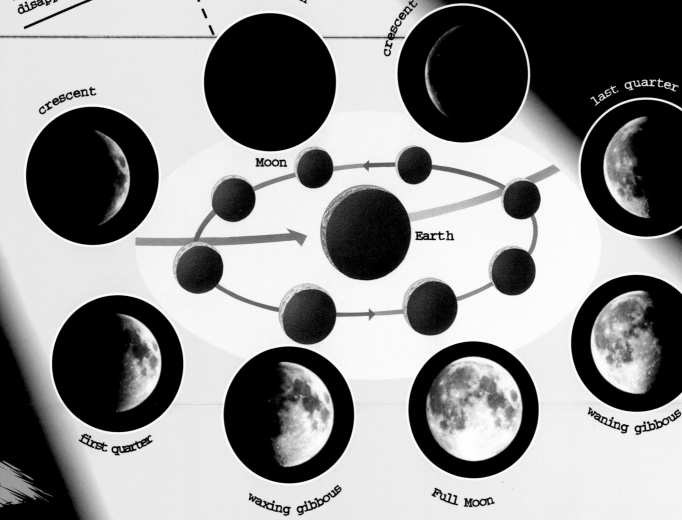

Sun

New Moon

crescent

crescent

last quarter

Moon

Earth

first quarter

waxing gibbous

Full Moon

waning gibbous

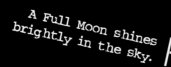
A Full Moon shines brightly in the sky.

Lunar eclipses

When Earth, the Sun, and Moon directly line up, there is an eclipse. A **lunar** eclipse (of the Moon) occurs when Earth lies between the Sun and a Full Moon, casting its shadow on the Moon. There are usually two or three lunar eclipses a year and they can be seen from anywhere on the half of Earth that is facing the Moon.

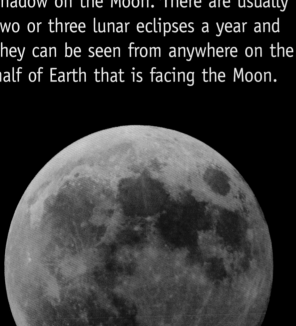
The Moon doesn't disappear during a lunar eclipse. Instead, it turns dark orange in color.

STAR FACT
The same half of the Moon is always facing toward Earth. The half that faces away from Earth is called the "far side."

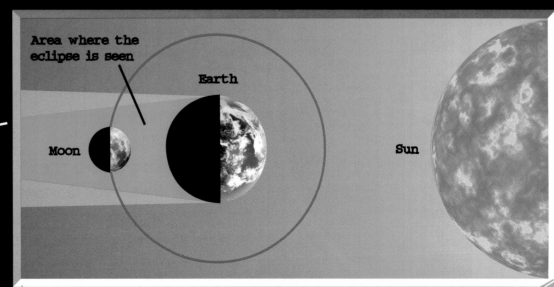

If the Moon passes through Earth's shadow when it is on the opposite side of Earth from the Sun, there is a lunar eclipse.

Area where the eclipse is seen

Earth

Moon

Sun

MARS, THE RED PLANET

There are huge dust storms

Although only half its size, Mars is very similar to Earth. Wispy clouds form a thin atmosphere and winds whip up clouds of dust in its desertlike landscape. A Martian day is just over 24 hours long. However, unlike Earth, Mars doesn't have any liquid water.

A rugged landscape

Mars is known as the red planet because it is covered in a layer of red soil that is rich in iron. Its surface is dry and rocky, and dust storms can sometimes cover the whole planet. It has not rained there for about three billion years.

Valles Marineris is named after Mariner 9, the Mars orbiter that discovered it in 1971. It is one of the biggest valley systems in the Solar System.

Vast valley

In one area of Mars, there was once a huge bulge on the surface. The bulge swelled, causing a deep crack in the crust. This crack is known as the Valles Marineris, a system of valleys that stretches a quarter of the way around the planet. It's long enough to fit across the United States.

Mega mountain

Mars has the largest mountain in our Solar System, called Olympus Mons. It stands at 16 miles high, which is three times higher than Mount Everest. Olympus Mons is a volcano, but has not erupted for millions of years.

Olympus Mons covers an area larger than the United Kingdom. It became so high after several volcanic eruptions in the same place.

Olympus Mons

Valles Marineris

Deismos

Phobos orbits Mars in 7.6 Earth hours and Deismos completes an orbit in 1.26 Earth hours. They both have uneven, cratered surfaces.

Phobos

Moons of Mars

Mars has two tiny Moons called Phobos and Deimos. They are probably asteroids that were pulled in by Mars' gravity. Phobos is the nearest to Mars and the largest at 17 miles across. Deimos is 10 miles across.

THE GIANT PLANETS

The four enormous planets on the edge of the Solar System—Jupiter, Saturn, Uranus, and Neptune—are very different from the rocky planets near the Sun. They are called "gas giants" because they are made from helium and hydrogen, which are the main gases that make up the Sun.

The big giants of the outer Solar System are Jupiter, Saturn, Uranus, and Neptune. All four planets have rings and moons.

Inside the giants

Both Jupiter and Saturn have layers of gas on the outside, making atmospheres with winds and clouds. Deep inside, hydrogen gas is squashed so much that it turns into liquid metal. All four gas giants have rocky cores about 12,500 miles across. Neptune and Uranus are blue in color because they have less hydrogen than Jupiter and Saturn, and more methane and ammonia.

rocky core

liquid metallic hydrogen

rocky core

Jupiter is mainly made of hydrogen. Clouds of frozen crystals float in its atmosphere.

Uranus' atmosphere contains a layer of methane clouds covered by haze.

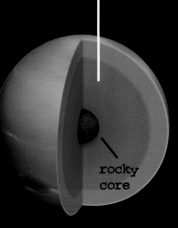

liquid rock, water, methane, and ammonia

Saturn's atmosphere is similar to Jupiter's. It is mainly made of liquid hydrogen.

rocky core

Neptune has bright clouds of frozen methane crystals high in its atmosphere.

rocky core

liquid metallic hydrogen

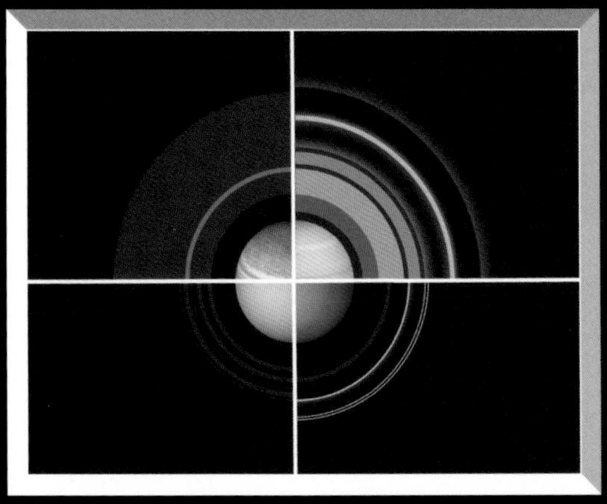

Rings of rock

Saturn has bright, wide rings made of chunks of ice. They are so bright that they can be seen from Earth. Uranus and Neptune both have narrow rings. Jupiter's rings are made up of very fine dust so can hardly be seen.

Jupiter

Saturn

Uranus

Neptune

This diagram shows the size of the rings. To make it easier to compare, the planets are shown at the same size.

Growing gas

When the Solar System formed, the gas giants grew much larger than the rocky planets because the outer Solar System was much colder. Nearer the Sun, icy material melted away, leaving only rock to form the planets. Farther out, clumps of icy material became part of the gas giants, and they grew bigger and bigger.

STAR FACT

It is impossible for spacecraft to land on the gas giants because there is no solid surface.

The growing giants attracted large amounts of gas when they formed.

Neptune

Saturn

Jupiter

Uranus

JUPITER, THE LARGEST PLANET

The largest planet in the Solar System, Jupiter is heavier than all the planets put together. Jupiter is the fastest-spinning planet, with a day lasting only 10 Earth hours. This speedy movement creates winds that travel at about 400 miles per hour.

Strong winds push light- and dark-colored bands of cloud around the planet.

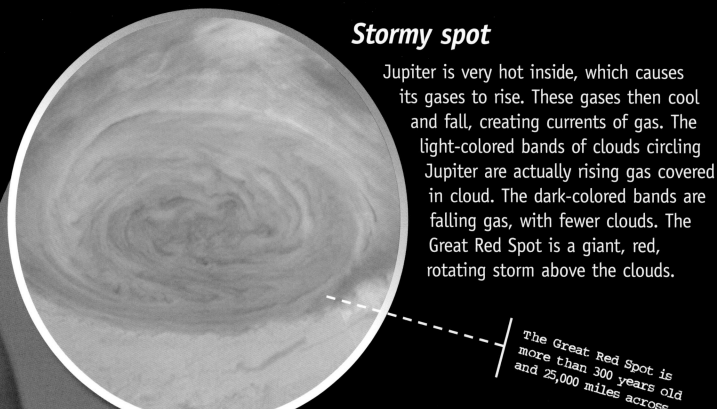

Stormy spot

Jupiter is very hot inside, which causes
its gases to rise. These gases then cool
and fall, creating currents of gas. The
light-colored bands of clouds circling
Jupiter are actually rising gas covered
in cloud. The dark-colored bands are
falling gas, with fewer clouds. The
Great Red Spot is a giant, red,
rotating storm above the clouds.

The Great Red Spot is
more than 300 years old
and 25,000 miles across.

Many moons

Jupiter has 67 known moons, including Ganymede, the
largest moon in the Solar System. The four moons
nearest to Jupiter are tiny. The next four moons are
large, called the Galilean moons. Much farther out
are two groups of four small, dark moons.

STAR FACT

Ganymede is 3,270 miles
across. That's larger than
the planet Mercury.

Io's surface is covered in
volcanoes. The moon looks
yellow in color because the
volcanoes spew out sulfur gas.

Part of
Ganymede's icy
surface is dark
and cratered,
and part of it
is smooth.

Europa is
covered by
smooth ice.

Callisto is made
mainly of ice and is
covered in craters.

FOCUS ON... GALILEO GALILEI

Italian mathematician and astronomer Galileo Galilei (1564–1642) was the first person to use a telescope for astronomy. He was also the first person to record his findings in The Starry Messenger, published in 1610.

Galileo started his career by studying medicine. He later swapped to mathematics.

STAR FACT

Galileo invented the pendulum clock after watching a bronze lamp swinging on a chain.

Galileo's observations

Through a telescope that Galileo made himself, he looked at the Moon, Venus, and Jupiter. In 1610, he discovered that there were four moons orbiting Jupiter. These moons are named after him—the Galilean moons. He also saw that Venus has phases, just like Earth's Moon.

The Galilean moons are Jupiter's four largest moons.

Changing ideas

Galileo's experiments proved that great thinkers of the past were wrong. Before Galileo, it was believed that the Moon was a smooth, perfect ball. He showed that it was actually rough and covered in mountains and craters. Another theory stated that heavier objects fall faster than lighter objects. To test this, Galileo rolled balls down slopes and worked out how fast they were traveling. He found that they all traveled at the same speed.

According to legend, Galileo dropped balls from the top of the Leaning Tower of Pisa, Italy, to test his idea.

SATURN, THE RINGED PLANET

The second largest planet in the Solar System, Saturn is easily recognized by the spectacular set of rings surrounding it. The planet is covered in fuzzy, yellow-colored bands of clouds, which form deep in the atmosphere, underneath a layer of haze.

Severe storms

Strong winds blow even harder than on Jupiter and are more than ten times faster than hurricane-force winds on Earth. About every 30 years, a giant storm forms. The last storm was in 2013, and swirling white clouds spread around the planet. They covered an area ten times the size of Earth.

During a storm in 1994, Saturn's winds reached a speed of 1,000 miles per hour.

Saturn has at least 62 moons, and all of them are made of rock and ice and covered in craters. The largest moon Titan is the only moon in our Solar System to have its own atmosphere, which is four times thicker than Earth's.

Herschel crater was formed on Mimas by an impact that nearly shattered the moon into pieces.

Many of the craters on Enceladus have been covered by ice.

Rings of ice

Saturn has thousands of individual, narrow rings, each made of millions of dirty chunks of ice, mixed with dust and rock. They may have formed when a moon or comet came too close to the planet, and was torn apart by its gravity.

Titan's atmosphere is mainly made of nitrogen, giving the moon an orange colored haze.

The smallest chunks of ice are only a few inches across, but the largest are about 0.5 miles across.

URANUS, THE TILTED PLANET

The seventh planet from the Sun, Uranus is covered in thick, blue clouds. It looks almost the same all over, with only very faint markings. Its icy atmosphere is so cold that the surface temperatures drop to a freezing -345 degrees Fahrenheit.

North Pole

Rolling around the Sun

Uranus is the only planet in the Solar System to spin on its side as it orbits the Sun. It is tilted at almost 98 degrees, meaning that its equator (center) and rings are vertical. It may have been knocked into this position after colliding with another planet.

At Uranus' north pole, there are 42 years of darkness, followed by 42 years where the Sun never sets.

STAR FACT

A day on Uranus lasts only 17 Earth hours.

Black rings

Uranus' rings were only discovered in 1977 when Uranus passed in front of a star and astronomers saw its brightness change. They realized that the star must have moved behind a set of rings. Uranus has nine narrow rings, made of black particles, making them almost invisible.

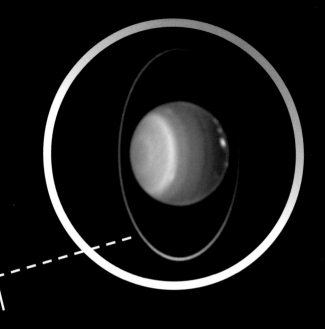

Uranus' rings reflect so little sunlight that they are invisible from Earth's surface. Only special space telescopes have taken photographs of them.

Shakespeare's moons

Uranus has 27 known moons, and many of them have been named after Shakespearean characters. The moons are made of ice and rock, and are covered in craters. Uranus's five largest moons also have long cracks and trenches.

Miranda's surface looks deeply scarred with craters and cracks.

Ariel is Uranus' brightest moon.

Titania is Uranus' largest moon, measuring 980 miles across.

Umbriel is covered in craters, measuring up to 130 miles across.

The surface of Oberon is slightly red in color.

FOCUS ON... WILLIAM HERSCHEL

William Herschel (1738–1822) was a professional German musician who liked to build telescopes. He became the King of England's astronomer and built many powerful telescopes, with which he discovered Uranus by accident.

Herschel was possibly the first astronomer to use the word "asteroid."

Discovering a new planet

Before Herschel's discovery, it was believed that there were only six planets—Earth, Mercury, Venus, Mars, Saturn, and Jupiter. In 1781, Herschel saw a dot of light through his telescope, which he believed was a comet or star. When he looked at it again a few days later, he realized that it must be a new planet.

STAR FACT

Herschel built more than 400 telescopes.

When Herschel discovered Uranus, it was the first planet that could not be seen with the naked eye.

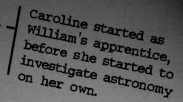

Uranus

Caroline started as William's apprentice, before she started to investigate astronomy on her own.

Keeping it in the family

Herschel worked with his sister Caroline (1750–1848). Together, they discovered Uranus. Caroline was also a great astronomer, identifying comets and producing star catalogs. Herschel's son John (1792–1871) cataloged all the stars in the Southern **Hemisphere**.

John Herschel followed in his father's footsteps, building telescopes and producing catalogs that contained thousands of stars.

NEPTUNE, THE WINDIEST PLANET

The fourth largest planet in our Solar System, Neptune is 31,763 miles across. Like Uranus, it is surrounded by blue clouds, made of extremely cold methane. Unlike Uranus, it also has white icy clouds.

High in Neptune's atmosphere, bright white clouds form. They are made of crystals of frozen methane gas.

Discovery

Neptune was discovered in 1845 by the English mathematicians John Couch Adams and Frenchman J. J. Leverrier. Both men were working independently at the Berlin Observatory in Germany, and they each worked out where the planet would be found. The orbit of Uranus took an unexpected turn, so they predicted that the gravity of another planet must be affecting it. Leverrier won the race to get his prediction checked.

Neptune has four thin, faint rings that are made of black dust.

Stormy weather

Winds around Neptune's equator blow at more than 1,250 miles per hour. They are the strongest known winds in the Solar System. The Great Dark Spot was a giant storm, seen in 1989 by Voyager 2. It has now disappeared.

The Great Dark Spot Storm was as large as Earth.

Unusual moons

Like other moons in the Solar System, Neptune's moons are made of ice and rock. The planet has 14 known moons. The large moons are named after characters from Greek myths. The two outer moons, Triton and Nereid, have unusual orbits. Triton goes around backward, and Nereid's orbit is the shape of a squashed egg.

Triton has an icy surface, covered in ridges, trenches, and craters. Plumes of ice erupt from below the surface, shooting 5 miles high.

STAR FACT

Triton is the coldest place in the Solar System, with temperatures dropping to -390 degrees Fahrenheit.

THE DWARF PLANETS

Beyond Neptune, hundreds of small objects orbit the Sun in a region called the Kuiper Belt. It was only discovered by astronomers in 1992.

Objects in the Kuiper Belt are often called Trans-Neptunian Objects (TNOs) because they are beyond Neptune.

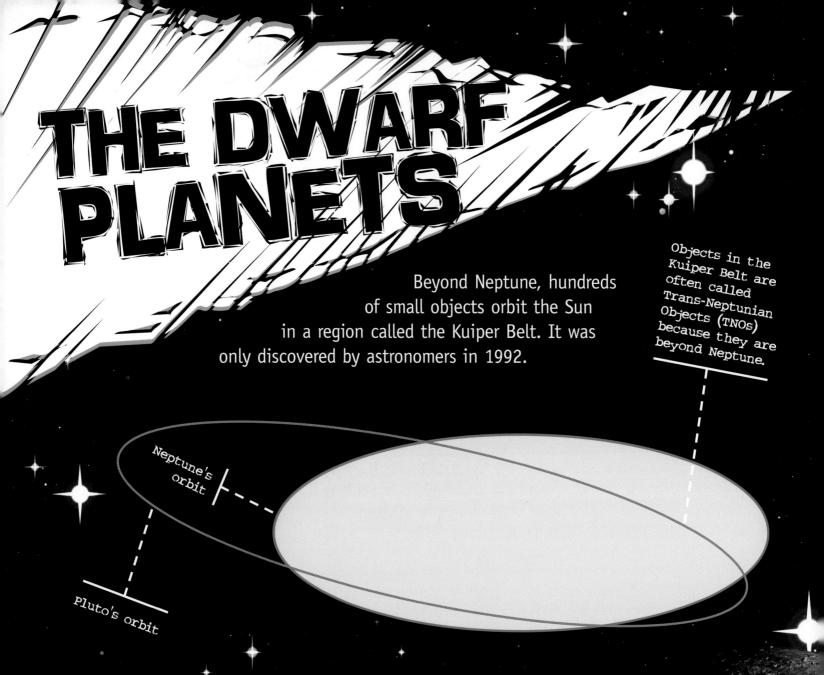

Neptune's orbit

Pluto's orbit

The first dwarf planets

In 2003, astronomers discovered a planet in the Kuiper Belt that was slightly bigger than Pluto. It was named Eris. After this discovery, in 2006, astronomers decided to class Eris and Pluto as dwarf planets because they are much smaller than the major planets of the Solar System.

It takes Eris 557 Earth years to complete one orbit of the Sun. Its moon is called Disnomia.

The list grows

Astronomers have found more than 1,000 objects in the Kuiper Belt, but they believe there could be around 70,000 objects. The five dwarf planets are Pluto, Eris, Ceres, Haumea, and Makemake.

Haumea is shaped like an egg and is one of the fastest-turning planets in the Solar System. It completes a spin in only four Earth hours.

Ceres is the only dwarf planet in the asteroid belt.

Astronomers have discovered frozen nitrogen on Makemake's surface. It take 310 Earth years for Makemake to orbit the Sun.

Disnomia

STAR FACT

Dwarf planets that orbit the Sun beyond Neptune are also called plutoids, named after Pluto.

PLUTO, THE DWARF PLANET

Pluto was once classed as the ninth planet in our Solar System, but it was extremely small in comparison to the other planets. In 2006, astronomers decided it should be called a dwarf planet instead.

Pluto is only 1,466 miles across.

Pluto's pole

layer of frost

rock and ice

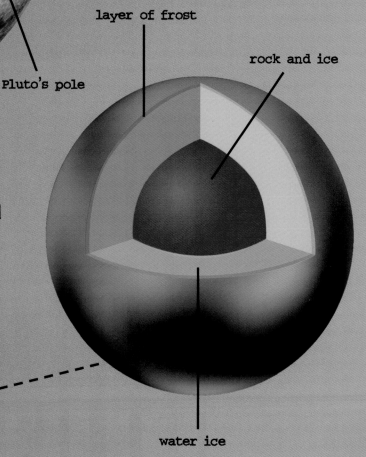

A ball of ice

Pluto is a rocky planet, which is smaller than Earth's Moon. It is probably made up of rock and ice inside, covered with a layer of watery ice about 200 miles thick. Its surface is covered by a thick layer of frozen gases—methane, carbon monoxide, and nitrogen—that is several miles deep.

Pluto is also covered in dark and light patches. One of the light areas is a bright cap on Pluto's pole.

water ice

On a different path

Pluto's orbit is shaped like an oval and it is tilted to 17 degrees. Sometimes there is a shorter distance between Pluto and the Sun than there is between Neptune and the Sun. It takes Pluto 248 years to complete one orbit of the Sun. A day on Pluto—the time it takes for the dwarf planet to spin around—lasts for 6.4 Earth days.

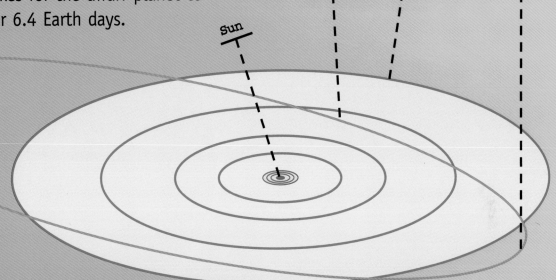

When close to the Sun, the ice starts to melt, which releases gases. This makes Pluto's atmosphere much thicker.

When farthest from the Sun, Pluto's atmosphere freezes into ice on the surface.

Neptune's orbit

Sun

Bulge or moon?

In 1978, astronomer James Christy noticed that Pluto appeared to have a bulge on one side. This was actually a large moon, later named Charon. The moon's surface is covered in frozen water. Pluto has five known moons: Charon, Styx, Nix, Kerberos, and Hydra.

Charon is half the size of Pluto, and its orbit is very close to the planet.

STAR FACT

Pluto was named by an 11-year-old girl from Oxford, UK.

ASTEROIDS, THE LEFTOVER ROCKS

Thousands of small, rocky chunks orbit the Sun. They are called asteroids, or the minor planets. They range in size from less than a mile across to more than 600 miles across. Most asteroids can be found in an area called the asteroid belt.

Vesta is one of the largest asteroids at 325 miles across. It was discovered in 1807.

Different materials

Asteroids were formed from leftover rock when the Solar System formed. They are not all made from the same material, however. Most of them are dark and dull, but some are light in color.

Asteroids are often covered in craters caused by constant collisions with smaller rocks.

The asteroid belt

An area between the orbits of Mars and Jupiter is called the asteroid belt. It contains thousands of objects. Scientists believe that the lumps of rock in the asteroid belt cannot clump together because the pull of Jupiter's gravity is too strong.

The asteroid belt forms a near-perfect donut-shaped ring around the Sun.

Mars

THE LARGEST ASTEROIDS

Name	Diameter across
Pallas	330 miles
Vesta	325 miles
Hygeia	265 miles
Davida	200 miles
Interamnia	200 miles
Cybele	175 miles
Europa	175 miles
Sylvia	170 miles
Patientia	170 miles

Jupiter

DIRTY SNOWBALLS

Bright comets look spectacular
as they whoosh across the sky.
Comets are balls of ice, followed by a
tail of streaming gas. About ten new comets are
discovered every year, but they are often very faint.
The brightest comets pass by every 10–20 years.

Heads and tails

Inside the head of a comet is a **nucleus** surrounded
by a huge cloud of gas and dust. This is called
the coma. As the comet travels nearer the
Sun, the icy nucleus starts to melt, creating
a tail of gas millions of miles long.

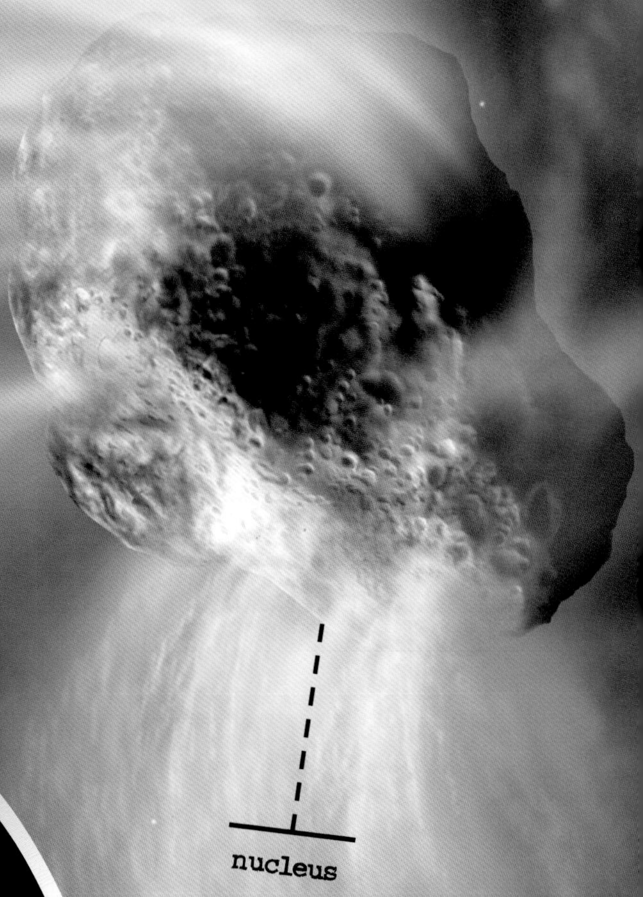

nucleus

Comet tails
always stream
away from the
Sun. The gas is
swept along by
the solar wind.

Comets have two
tails—a curved
yellow dust tail
and a straight
blue gas tail

Near and far

Most comets orbit the Sun then
disappear into space for thousands of years.
They gradually move across the sky like planets and
are usually visible for a few weeks. The closest a comet has
ever come to Earth was in 1770, when Lexell's Comet came
within 1.4 million miles.

FAMOUS COMETS

Most bright comets arrive unexpectedly and are only seen once. However, some comets pass by regularly. New comets are often named after their discoverer. The most famous comet is Halley's Comet, named after Edmond Halley.

In 1682, Edmond Halley realized that the comet he saw was the same one that had been seen 76 years earlier. Halley's Comet has been observed every 76 years for 2,200 years! It will next appear in 2062.

STAR FACT

Halley's Comet was shown on the Bayeux Tapestry, a cloth showing the Battle of Hastings in 1066.

Millions of people around the world saw Comet Hale-Bopp in 1997, and it is one of the brightest comets ever seen. Its icy nucleus is 25 miles across. That's twice as big as Halley's Comet.

Comet Hale-Bopp

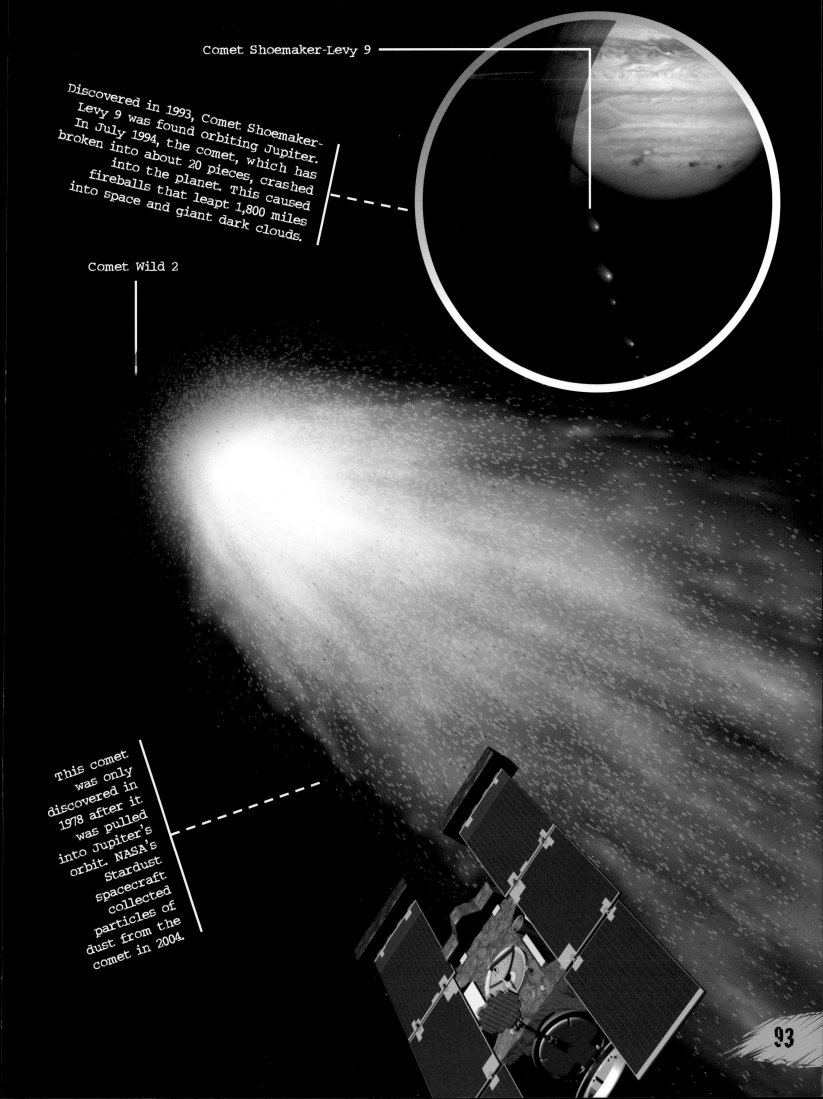

Comet Shoemaker-Levy 9

Discovered in 1993, Comet Shoemaker-Levy 9 was found orbiting Jupiter. In July 1994, the comet, which has broken into about 20 pieces, crashed into the planet. This caused fireballs that leapt 1,800 miles into space and giant dark clouds.

Comet Wild 2

This comet was only discovered in 1978 after it was pulled into Jupiter's orbit. NASA's Stardust spacecraft collected particles of dust from the comet in 2004.

METEORS AND METEORITES

The Solar System is full of dust and rocks. When dust particles crash into Earth's atmosphere, they burn up, leaving a fiery trail. This fast-moving "shooting star" is called a **meteor**. Large space rocks that travel through Earth's atmosphere and reach the ground are called meteorites.

A meteorite hit Earth about 50,000 years ago in Arizona. It measured 4,000 feet across and weighed 10,000 tons.

STAR FACT

Some scientists believe that the dinosaurs were wiped out when a comet hit Earth about 65 million years ago.

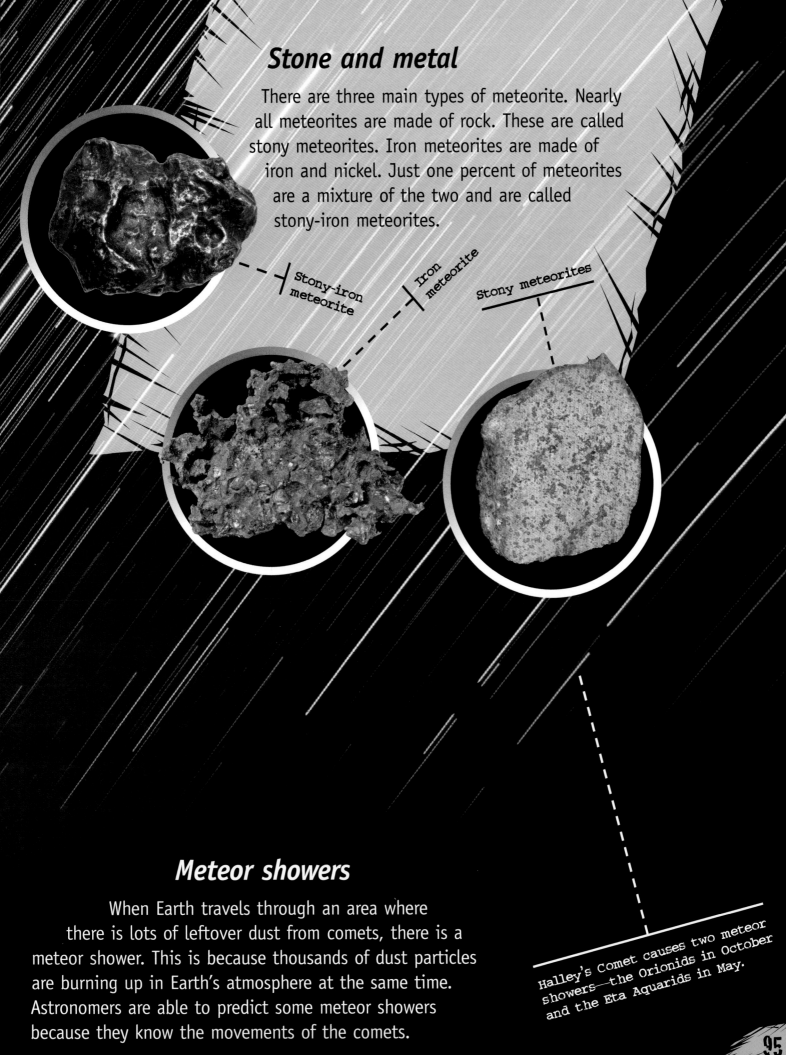

Stone and metal

There are three main types of meteorite. Nearly all meteorites are made of rock. These are called stony meteorites. Iron meteorites are made of iron and nickel. Just one percent of meteorites are a mixture of the two and are called stony-iron meteorites.

Stony-iron meteorite

Iron meteorite

Stony meteorites

Meteor showers

When Earth travels through an area where there is lots of leftover dust from comets, there is a meteor shower. This is because thousands of dust particles are burning up in Earth's atmosphere at the same time. Astronomers are able to predict some meteor showers because they know the movements of the comets.

Halley's Comet causes two meteor showers—the Orionids in October and the Eta Aquarids in May.

WHAT IS SPACE EXPLORATION?

Scientists explore space so that they can understand how the Universe works and what is "out there." This also allows them to understand Earth better. We started to explore space in the 1940s. Since then there have been many missions, including sending astronauts to the Moon, visiting other planets with unmanned spacecraft, and even fly-bys of the Sun.

Missions in space

Spacecraft have visited the Sun, Moon, planets, comets, and asteroids. They carry lots of special equipment such as cameras to collect information to send back to Earth. Some have even collected space rocks to be examined by scientists. This helps them to see if there is life on another planet.

The Dawn spacecraft was launched in 1997. It will travel to the asteroid Vesta and dwarf planet Ceres. It will reach Ceres in 2015.

STAR FACT

Virgin Galactic plans to take "space tourists" into space in the future. The tickets cost $200,000 each.

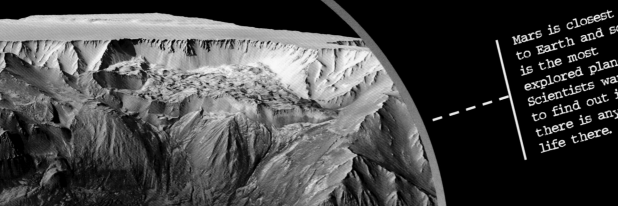

Mars is closest to Earth and so is the most explored planet. Scientists want to find out if there is any life there.

Types of spacecraft

Different types of spacecraft are used for different missions. Rockets take **satellites** and other crafts into space. In a fly-by, a spacecraft passes close to an object. An orbiter circles a planet or moon. A lander gathers information from the surface of an object. A probe is dropped into an atmosphere, and takes measurements as it falls to the surface.

EARLY EXPLORERS

In 1957, Russia changed history forever. It launched the first-ever satellite into space. This marked the start of the "Space Age." The satellite, *Sputnik-1*, was the size of a beach ball and took 98 minutes to orbit Earth. It collected information about Earth's atmosphere. It spent three months in space before burning up in the Earth's atmosphere.

Sputnik-1 traveled 37 million miles during its space mission.

Early Moon missions

The first missions to the Moon were made by Russia's *Luna 1* and *Luna 2* spacecrafts, both launched in 1959. *Luna 1* completed a fly-by of the Moon and discovered the solar wind. *Luna 2* was launched eight months later and deliberately crashed into the Moon. It took measurements of the Moon's gravity.

Luna 2 was the first spacecraft in the world to reach another space object.

Man in space

The first human to travel into space was Russian cosmonaut Yuri Gagarin. In 1961, he completed one orbit of Earth in **Vostok 1.** The spacecraft contained radios, a life support system, and an ejection seat. Gagarin spent 108 minutes in orbit and traveled 203 miles before reentering Earth's atmosphere.

STAR FACT

The first animal in space was a dog named Laika, on Russia's Sputnik-2 in 1957.

Only the manned module of *Vostok 1* came back to Earth. Gagarin ejected from the spacecraft and landed by parachute.

ROCKETING INTO SPACE

To go into space and stay above Earth, a spacecraft needs to reach a speed of 17,500 miles per hour. A rocket is the only vehicle that can travel this fast. The fastest planes only get to one-tenth of this speed. Once a rocket has released its spacecraft, it falls back to Earth, usually landing in the ocean.

Launch power

Rockets need a huge amount of power to overcome Earth's gravity. To do this, they burn a propellant. This is made of fuel and an oxidizer, which helps the fuel to burn in space, where there is no oxygen. The hot gases that are produced push the rocket upward.

NASA's *Atlas V* rocket is 191 feet high. It was first launched in 2002, and has completed several successful missions since then.

A rocket is actually made up of 2–3 booster rockets, each with its own engine. When each booster rocket uses up its fuel, it falls away. The final rocket launches into space.

booster rocket

Carrying cargo

Spacecraft and other cargo that a rocket carries are called its payload. Rockets usually travel into Earth's orbit, before the nose cone opens and releases the spacecraft. Saturn V carried the heaviest payloads— up to 45 tons to the Moon and 120 tons into Earth's orbit.

STAR FACT

The first liquid-fuel rocket was launched in 1926 by Robert Goddard. It stayed in the air for 2.5 seconds and climbed to a height of 39 feet.

Saturn V launched the Apollo 11 spacecraft into space on July 16, 1969, for the first Moon landing.

THE SPACE SHUTTLE

The space shuttle was a reusable spacecraft used for manned missions. It was first launched in 1982, and completed 135 missions before it was retired in 2011. It launched probes and satellites and took crew members to the International Space Station.

thrusters

cargo bay with satellite and instruments

The space shuttle landed on Earth by gliding onto a runway. The pilot used a parachute and brakes to slow down the Shuttle.

robot arm

Rocket boost

The space shuttle was the only vehicle that acted like a rocket and launched itself into space. It had three rocket thrusters and two solid-fuel boosters, which gave it almost as much power on launch as the *Saturn V* rocket.

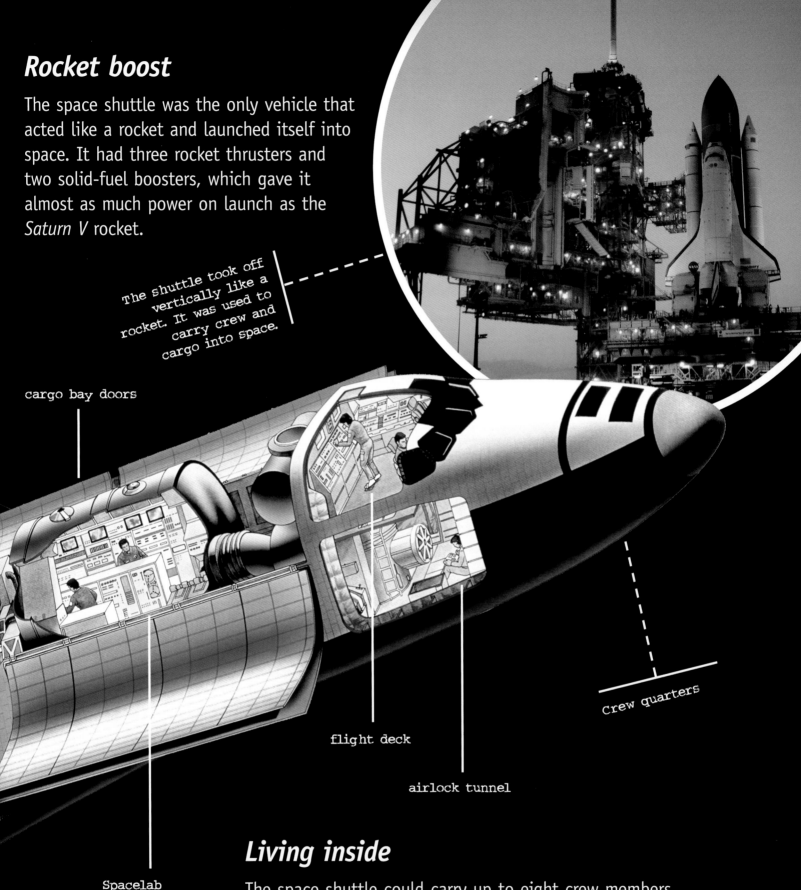

The shuttle took off vertically like a rocket. It was used to carry crew and cargo into space.

cargo bay doors

flight deck

airlock tunnel

Crew quarters

Spacelab

Living inside

The space shuttle could carry up to eight crew members. The crew had to live on the spacecraft for around seven days, so it came well equipped. It had a flight deck where the pilot controlled the craft. The crew quarters included beds and eating areas. The onboard laboratory, called Spacelab, allowed the crew to carry out experiments. The crew used the robot arm to grab satellites in need of repair.

ASTRONAUTS

Astronauts are people who travel in spacecraft. Space is a dangerous place so astronauts spend months training for their missions. They need to be able to walk in a weightless atmosphere, cope with incredibly high and low temperatures, and operate the spacecraft.

Weightlessness causes astronauts to grow a few inches during a mission.

STAR FACT

Once a NASA astronaut has flown in space, they are awarded a gold lapel pin.

Suited up

An astronaut's spacesuit is very important because it protects them from harmful rays from the Sun. It covers every part of the astronaut's skin. A spacesuit has 14 layers to make it strong. It also holds oxygen so the astronaut can breathe and water to keep the astronaut cool in high temperatures. It is also waterproof and fire resistant.

A spacesuit contains everything that an astronaut needs to survive.

In training

NASA astronauts train for up to two years. During this time, they are exposed to as many space conditions as possible. As well as flying spacecraft simulators, astronauts train on plunging jet aircraft to make sure that they can cope with the G-force of a space flight.

Astronauts train in special water tanks. Walking in water is similar to moving in space.

SPACE STATIONS

The *Soyuz* spacecraft carried astronauts from Earth to the space station. The *Soyuz* 11 crew lived on the space station for 24 days.

Space stations are large spacecraft where astronauts can live and work for several months at a time. They are vital for us to understand more about space, and they allow scientists to carry out research in space itself, with conditions that are not found on Earth. There have been five space stations, including the International Space Station.

Ahead of the game

The first space station, *Salyut 1*, was launched in 1971 by Russia. It measured 45 feet in length, and consisted of three main areas. These areas contained sleeping and dining areas, storage for food and water, and control stations. It stayed in space for only five months.

STAR FACT

China launched its first space station in 2011. It is called Tiangong-1 and in June 2012, three crew members successfully went on board.

First US station

The USA's first space station was called *Skylab* and was placed into orbit in 1973. During the launch, the space station was damaged, so the first crew needed to repair it. From its launch until 1974, three crews spent a total of 171 days onboard the space station.

Skylab was built to test the effects of space conditions on the human body over a long period of time.

Mir was badly damaged in a collision with a supply vehicle in 1997, but continued to work.

Staying in space

The world's first permanent space station—Mir—was launched by Russia in 1986. It was built in stages, with new modules added over the years. During its service until 2000, it made 89,000 orbits of Earth. The space station had living quarters, docking ports, and laboratories for studying Earth, stars, and other galaxies.

INSIDE THE ISS

The International Space Station is a giant space station that was built in space. Its first piece was launched in 1998. After 13 years, the ISS was finally completed in 2011. It is a joint project between the United States and 14 other countries. The first three crew members boarded the station in 2000, and lived there for five months. It has been manned ever since.

Reaching the ISS

It takes two days for a spacecraft to reach the ISS from Earth. When it arrives at the ISS, it must slowly pull alongside at the docking station. After about 30 minutes, the hatch opens and the astronauts can board the station. There is a permanent crew of six people.

On a clear night, the ISS is visible in the sky using the naked eye.

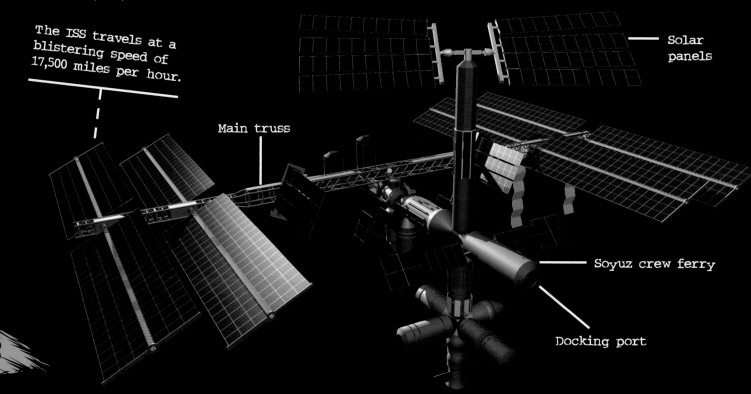

The ISS travels at a blistering speed of 17,500 miles per hour.

Solar panels

Main truss

Soyuz crew ferry

Docking port

Living with no weight

Inside the space station, astronauts are weightless. They need to use exercise machines to keep their bones and muscles strong. Daily tasks such as washing can take them a long time. There is no gravity, so everything floats. Astronauts need to be careful when using water, so they use small amounts at a time. If they were to spill any, it would float and could end up short-circuiting the onboard-computers.

Everything needs to be strapped down on the space station—even the astronauts when they are sleeping!

EYE IN THE SKY

Manmade satellites are important because they help scientists to study Earth and other objects in the Solar System. There are hundreds of satellites in space, launched into orbit by rockets. They can collect data quickly because they have a much wider view than anything on Earth.

What's the weather?

Some satellites monitor the weather. They can see how clouds are forming and moving, where the wind and rain are, and measure the temperature of the air and ground. They also take pictures of hurricanes and storms from above.

Satellites help meteorologists predict the weather and track storms such as hurricanes.

Messages to the world

Satellites are used for communications, sending television and telephone signals around the world. Radio signals are sent up to a satellite and then instantly beamed down to an antenna on your house. Before communications satellites, long-distance phone calls were not possible.

There are also about 30 GPS satellites in space, which allow people to locate their exact position using a GPS locator.

When Iceland's Eyjafjallajökull volcano erupted in March 2010, satellites were able to monitor the movement of the ash plume.

STAR FACT

In 2009, two communications satellites crashed into each other. This is the first time manmade satellites have collided.

Studying Earth

Satellites that monitor Earth collect information about the land and oceans, as well as measure the levels of gas such as carbon dioxide in the atmosphere. They are also used to track disasters, such as wildfires and volcanic eruptions.

MISSIONS TO THE MOON

Only the Apollo missions have taken humans to another "world." From 1969 to 1972, there were 11 crewed missions. Only six Apollo spacecraft have landed on the Moon, each with two astronauts. The most famous mission was in 1969 when astronauts first stepped on the Moon's surface.

The Lunar Module landed on the Moon with the astronauts inside.

After the mission, just before entering Earth's atmosphere, the Command Module (containing the astronauts) separated. It landed in the ocean.

The Apollo spacecraft has three main parts— the Command Module, Service Module, and Lunar Module.

Lunar Module

Command Module

Service Module

It took three days for the space to reach the Moon. During this time, the astronauts were squeezed into a small space where they had little room to move.

Journey to the Moon

The Apollo spacecraft were launched from Earth on *Saturn V* rockets, which were the largest rockets ever built. *Saturn V* needed three rocket stages to give them enough power to thrust its enormous payload into space toward the Moon.

STAR FACT

During the Apollo missions, 842 pounds of Moon rocks were collected.

THE APOLLO MISSIONS

Mission	Date of Launch	Highlights
Apollo 11	July 1969	The first humans to walk on the Moon were Neil Armstrong and Buzz Aldrin
Apollo 12	November 1969	The aim of the mission was to complete a more exact landing
Apollo 13	April 1970	The spacecraft did not land on the Moon due to an explosion, but managed to take photographs instead
Apollo 14	January 1971	Alan Shepard traveled a distance of 9,000 feet across the Moon's surface
Apollo 15	July 1971	First use of a Moon buggy
Apollo 16	April 1972	A successful mission to explore the Moon's highlands
Apollo 17	December 1972	The first night launch of a manned space mission

Exploration

Astronauts spent 3–4 days on the Moon, collecting data about the particular area they landed on. They gathered space rocks and dust, carried out experiments, and explored the terrain.

The last three Apollo missions have used a Moon buggy for astronauts to explore a much wider area.

WALKING ON THE MOON

The Apollo 11 mission, launched in July 1969, is the most famous because it was the first time humans had walked on the Moon. The first astronauts to walk on the surface were Neil Armstrong and Buzz Aldrin. Michael Collins stayed on the Command Module.

The Apollo 11 mission lasted for eight days.

A tricky landing

As the Landing Module neared the Moon's surface, the astronauts realized that they were heading toward an area covered in boulders. Armstrong took control of the module, and with Aldrin's guidance, he avoided hitting the rocks. When they landed, the craft only had 25 seconds of fuel left, seconds away from aborting the mission.

A few minutes after stepping on the Moon, Aldrin began collecting soil samples.

Famous words

President J. F. Kennedy set the goal of the Apollo 11 mission in 1961—to land on the Moon and then return to Earth. Images of the landing were shown all over the world. It is estimated that 530 million people heard Neil Armstrong's famous words, "One small step for man, one giant leap for mankind." The mission changed history, proving that humans could survive in another world.

Neil Armstrong stepped onto the Moon 109 hours and 42 minutes after launch. The Apollo 11 astronauts planted a United States flag in the Moon's surface.

Neil Armstrong's imprint on the Moon.

VISITING THE ROCKY PLANETS

Mercury, Venus and Mars are the closest planets to Earth, making them easier to explore than the distant outer planets. The first visit to a rocky planet was in 1961 when Russian *Venera 1* completed a flyby of Venus.

Russian probe *Venera 1* had an 8ft antenna to transmit radio waves back to Earth.

Orbiting Mercury

Since arriving in March 2011, the *Messenger* spacecraft completed over 3,000 orbits of Mercury. It beamed back thousands of images of the surface, revealing craters and huge basins. For the first time, the entire surface of Mercury has been mapped. In 2015, after *Messenger's* fuel ran out, it crashed into Mercury, ending its mission.

Messenger returned thousands of photographs of Mercury, showing the planet's surface in sharp detail.

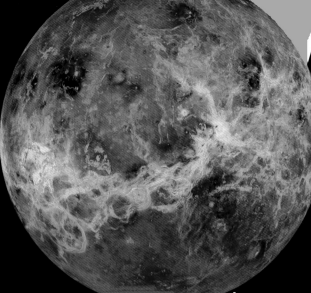

he Magellan mission

e surface of Venus is hidden by clouds. In 1989, the
gellan spacecraft orbited the planet to make a map
its surface using radar. This used radio waves, which
sily passed through the clouds. A computer then
rked out what the surface of Venus looked like.

Venus

Each time *Magellan* orbited the planet, it mapped a strip 15 miles wide and 6,000 miles long. It took three years to build a complete picture of Venus.

MISSIONS TO THE ROCKY PLANETS

Planet	Spacecraft	Date	Highlights
Mars	Mariner 4	1964	First fly-by of Mars. Sent back 22 photographs showing its cratered surface
Mars	Mariner 9	1971	First spacecraft to orbit another planet. It returned detailed images of the surface of Mars
Mars	Viking 1	1975	First successful lander on Mars. It returned images for six years
Mars	Mars Pathfinder	1996	First US robotic exploration of another planet's surface
Mars	Spirit and Opportunity	2004	Rovers with equipment to monitor Mars, such as cameras and microscopes
Mars	Reconaissance Orbiter	2006	Sharpest view of the surface of Mars.
Mars	Curiosity	2012	Aim of this rover is to explore Mars for evidence of bacterial life
Mercury and Venus	Mariner 10	1974-1975	First spacecraft to visit Mercury
Mercury	Messenger	2011	First spacecraft to orbit Mercury
Venus	Venera 1	1961	First fly-by
Venus	Venera 3	1965	First object launched from the Earth to reach another planet. It crash-landed on Venus
Venus	Venera 4	1967	First probe to return data on the atmosphere and surface of Venus
Venus	Venera 7	1970	First successful landing on another planet
Venus	Pioneer Venus	1978	Explored the atmosphere of Venus
Venus	Magellan	1989	Completed first detailed mapping of another planet's surface by radar
Venus	Venus Express	2006	Mapping the surface and atmosphere of Venus

EXPLORING MARS

Aside from Earth, Mars is the most explored planet in the Solar System. The first spacecraft to complete a fly-by was *Mariner 4* in 1965. Since then, there have been more than 40 missions in total, but only a few have been successful.

Viking orbiter

Looking for life

Several missions have been launched to search for primitive life on Mars, such as bacteria. Some spacecraft have an orbiter and lander. The orbiters use cameras to map the planet in detail. The landers test the soil and rocks. So far no evidence of life has been discovered.

Viking 1 and *Viking 2* returned images to Earth for several years after landing.

Viking lander

Touring the surface

One of the amazing ways scientists are exploring Mars today is by using robot vehicles to tour around the planet. These unmanned rovers are launched on spacecraft from Earth, carefully placed on Mars, and then controlled by radio signals from Earth. A little like remote-control cars!

The *Pathfinder* mission lasted for three months. It collected an amazing 2.3 billion pieces of information, including more than 17,000 images.

Pathfinder image of Mars

Sojourner rover

Curiouser and curiouser

Since 1996, four different rovers have landed on Mars to explore the planet's soil and rocks. The most recent mission involves a car-sized rover called *Curiosity*. On August 5, 2012, *Curiosity* was lowered onto the surface of Mars with parachutes, rockets, and cranes. The rover works like a moving laboratory, full of scientific equipment. It has 17 high-tech cameras, a scoop to gather and study dust, and it even has a laser to blast and drill rocks. The data beamed back by *Curiosity* will help us learn if Mars could have supported life forms in the past, or even today.

Curiosity is still working on Mars. It is hoped that the data *Curiosity* collects will help any future manned exploration.

360 degree panorama image of Mars taken by *Curiosity*.

VISITING THE GIANTS

The Voyager mission has made many great discoveries, including rings and moons. *Voyager 2* is the only craft to have visited Neptune and Venus.

There have been fewer missions to the gas giants because they are much farther away. It takes at least two years for a craft to reach Jupiter. Spacecraft cannot land on a gas giant's surface, so scientists use flybys and probes to collect data.

Touring the giants

Launched in 1977, *Voyager 2* is the only craft to have completed a flyby of the four outer planets. It reached Jupiter in 1979, Saturn in 1981, Uranus in 1986, and Neptune in 1989. Both *Voyager 1* and *Voyager 2* will continue to take measurements in space, until 2020.

Voyager 2 image of Neptune

Galileo spacecraft

In the clouds

In 1989, the *Galileo* spacecraft started on its six-year journey to Jupiter. In 1995, it dropped a probe into Jupiter's atmosphere. This orbiter had six instruments that monitored Jupiter for 57 minutes while it descended through the clouds. It took close-up photographs of Jupiter and its moons. These are the best that have ever been taken.

The probe traveled at a speed of more than 106,000 miles per hour when it entered Jupiter's atmosphere. It was slowed down by a parachute.

To the ringed planet

In 2004, the *Cassini-Huygens* spacecraft arrived at Saturn, after a seven-year journey. *Cassini* is the orbiter and *Huygens* was the probe that landed on Saturn's moon Titan in 2005. They have both provided vital information about the planet, its rings, and moons. Cassini will continue to orbit Saturn until 2017.

STAR FACT

More than 5,000 people have worked on the Cassini-Huygens mission. In total, the mission has cost more than $3.27 billion.

Cassini–Huygens spacecraft

Saturn

On Titan, the Huygens probe discovered huge oily lakes and enormous sand dunes created by the wind.

probe

121

JOURNEY TO THE SUN

There have been few missions to the Sun. This is because the spacecraft must be able to withstand enormous heat—9,941 degrees Fahrenheit—hot enough to melt metal, as well as safely travel 92.9 million miles.

Helios 1 and *Helios 2* were launched from the Kennedy Space Center in Cape Canaveral. They set a speed record for spacecraft of 157,078 miles per hour.

Speeding to the Sun

In the 1970s, *Helios 2* was one of two probes launched to orbit the Sun. *Helios 2* came within 27 million miles of the fiery ball, 2 million miles closer than its twin craft. It collected information on the Sun until 1985.

The Sun's wind bubble

Launched in 1990, *Ulysses* was the first spacecraft to explore the regions of space over the Sun's **poles**. It was sent into Jupiter's gravity and then pushed into an orbit that passed over the Sun. It monitored the heliosphere, a huge bubble in space created by the solar winds.

Ulysses passed the Sun 185 million miles above the poles.

Storm watcher

The Solar and Heliospheric Observatory (SOHO) was launched in 1995 to watch the Sun and monitor the raging storms on its surface. It returns images and data on the Sun's atmosphere, surface, and hot core. SOHO has also discovered thousands of comets. Its studies help scientists to predict and monitor space events that may affect Earth.

SOHO has taken special photographs that show the temperature of the Sun and its corona.

WHAT IS ASTRONOMY?

Observing the night sky to study stars, planets, moons, and galaxies is called astronomy. Astronomers use telescopes to see space objects that are too far away to be seen with the naked eye. Anyone can look into the sky to discover what's out there.

The eyepiece of the telescope magnifies the image, so you can see sights such as the Moon's craters clearly.

STAR FACT

Stars appear to twinkle because their light is disturbed by the movement of air in the Earth's atmosphere.

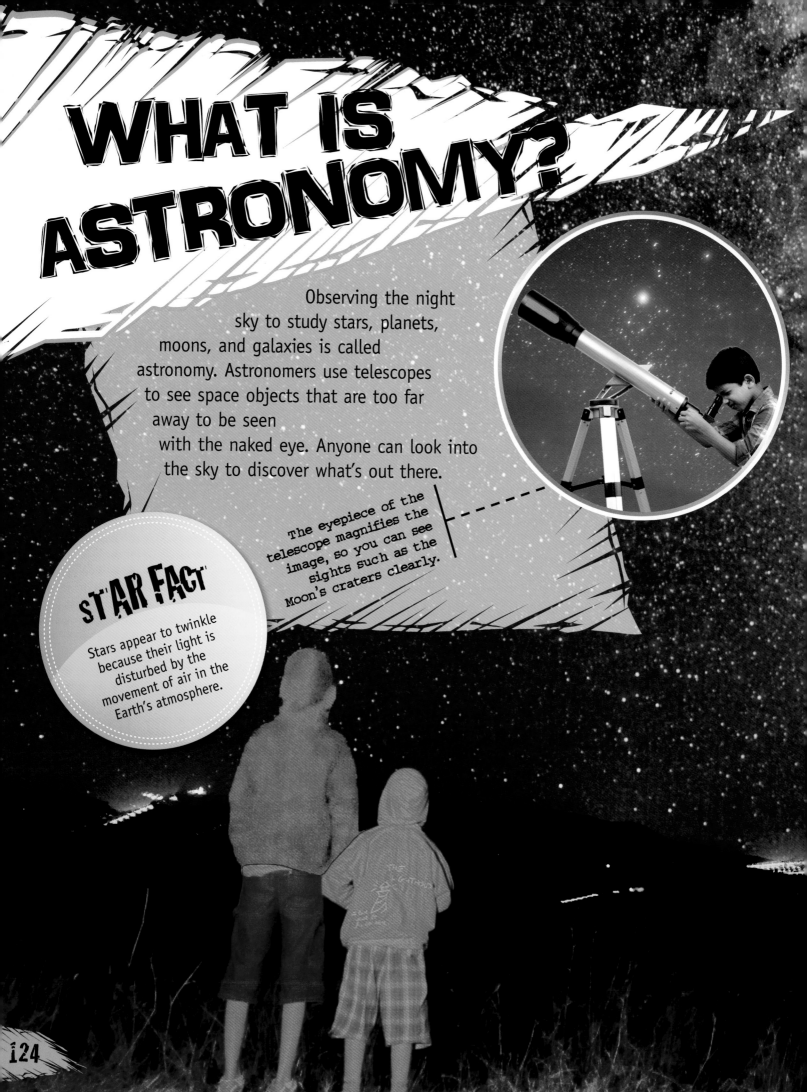

Starting out

To see the stars well, you should be away from the bright lights of a city or town. In a dark area, you may be able to see ten times as many stars. When your eyes get used to the dark, you will start to see more and more bright objects in the night sky.

The best time to watch the stars is on a clear, still night when there are no clouds in the sky.

A reflecting telescope uses a lens.

Magnifying the sky

Binoculars and telescopes are important pieces of equipment for skywatching. A telescope uses a lens or curved mirror to collect and focus light into your eye. As the lens is much larger than the pupil in your eye, it can catch more light, so it can bring into view faint stars and galaxies.

Binoculars aren't as powerful as telescopes, but they are less expensive and a good starting point.

A refracting telescope uses a curved mirror.

125

STAR MAPS

Since ancient times, people have recognized patterns of stars, and have named them. These patterns of stars are called constellations, and many names used today were given by Greek and Arabic astronomers thousands of years ago. Today, astronomers use 88 constellations to cover every part of the night sky.

Naming the stars

Most of the brightest stars in the sky were named by Arabic astronomers more than 1,000 years ago. Some bright stars are also named after their constellations. In the 17th century, the brightest stars were given a Greek letter, such as Alpha, followed by their constellation names. So some stars are known as Alpha Orionis and Beta Orionis. Faint stars are given only numbers.

Modern star maps are used by amateur observers. The lines joining the bright stars show the pattern of the constellation. This helps amateur observers to recognize the constellation in the sky.

The zodiac

Twelve constellations lie in a belt in the sky called the zodiac. During the day, we cannot see the stars because the Sun's light is too strong. Month by month, the Sun follows the zodiac belt, across the constellations. People once thought it was a magical pathway through the stars.

The zodiac belt includes animal constellations such as Leo (the Lion) and Cancer (the Crab). Only one constellation isn't related to animals—Libra (the Scales).

Capricorn
(the Sea Goat)

Sagittarius
(the Centaur,
or Archer)

Aquarius
(the Water-bearer)

AUGUST

JULY

SEPTEMBER

Scorpio
(the Scorpion)

Line of sight

Pisces
(the Fish)

JUNE

OCTOBER

Libra
(the Scales)

MAY

NOVEMBER

Aries
(the Ram)

Virgo
(the Maiden)

DECEMBER

APRIL

Taurus
(the Bull)

JANUARY

MARCH

FEBRUARY

Leo
(the Lion)

Gemini
(the Twins)

Cancer
(the Crab)

When the sun is in line with Aquarius, we say the sun is in Aquarius.

GAZING AT THE STARS

The stars rise in the east and set in the west, just like the Sun. The stars seem to move across the night sky because Earth is spinning. It takes 23 hours and 56 minutes for a star to return to exactly the same position in the sky. This is 4 minutes earlier than it takes the Sun each day.

Sirius (the Great Dog) is the brightest star in the night sky. It is 8.5 light years away.

How bright?

A star's brightness in the night sky is called its magnitude. The brighter a star is, the lower its magnitude. The brightest stars have a magnitude of -1 or less. The magnitude system was invented by Greek astronomer Hipparchus. In 120 BC, he divided the stars into six groups of brightness, simply using the naked eye.

Always shining

Some constellations stay above the horizon, so they never rise or set. They are called circumpolar constellations because they seem to circle around the north or south poles. In Canada and northern USA, the Big Dipper in the Great Bear constellation is circumpolar. In North America and Europe, the Little Bear is circumpolar.

This photograph is taken with a very long exposure. As Earth turns, the stars seem to circle around the North Pole, leaving a trail

STAR FACT

The bright star above the north pole is called Polaris, or the North Star. It hardly seems to move in the night sky.

TOP TEN BRIGHTEST STARS

Name	Constellation	Magnitude
Sirius	Canis Major (the Great Dog)	-1.5
Canopus	Carina (the Keel)	-0.7
Alpha Centauri	Centaurus (the Centaur)	-0.3
Arcturus	Bootes (the Herdsman)	0.0
Vega	Lyra (the Lyre)	0.0
Capella	Auriga (the Charloteer)	0.1
Rigel	Orion (the Hunter)	0.1
Procyon	Canis Minor (the Little Dog)	0.4
Achernar	Eridanus (the River)	0.5
Betelgeuse	Orion (the Hunter)	0.5

HOW TO USE STAR MAPS

Star maps show the brightest stars and most well-known constellations. They help astronomers to find their way around the night sky. Once you find one constellation, a star map shows you what is nearby.

Polaris

Big Dipper

Near Polaris is a famous pattern of stars called the Big Dipper (or Plow).

Where in the world?

The stars that you can see in a clear sky depend on three things.
1. Your **latitude** or how far north or south of the equator you are
2. The time of year
3. The time of night

The equator is an imaginary line around the center of the Earth

North America

Europe

Asia

Africa

South America

Australia

Antarctica

The Miller
PLANISPHERE
Puts the stars within everyone's reach
Latitude 40° North

Clever charts

A planisphere is a special circular chart that shows you exactly which stars are visible in the night sky at a certain time on a certain date. It is made of two discs, one fixed on top of the other. The bottom disc is a star chart with the months marked around the edge. The top disc has a hole in the center. The hours of the day are marked around the edge.

To use the planisphere, you line up the date and time, and the stars that are visible in the night sky are shown in the hole.

STARS IN THE NORTHERN HEMISPHERE

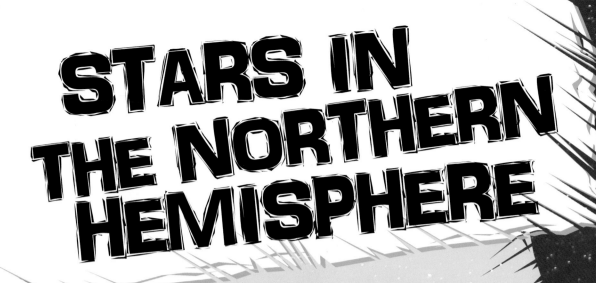

If you live north of the equator,
you live in the Northern Hemisphere.
Constellations in the sky seem to move across
in a counterclockwise direction.

Can you guess
which
constellation
this is?

STAR FACT

Orion is visible in the Northern and Southern hemispheres.

Orion (Hunter)

Gemini (Twins)

Taurus (Bull)

Cancer (Crab)

Pisces (Fish)

Perseus (Hero)

Auriga (Charioteer)

Andromeda (Chained Princess)

Leo (Lion)

Cassiopeia (Queen)

Polaris (Pole Star)

Ursa Major (Great Bear)

Cepheus (King)

Pegasus (Winged Horse)

Corona Borealis (Berenice's Hair)

Cygnus (Swan)

Lyra (Lyre)

Hercules (Hercules)

STARS IN THE SOUTHERN HEMISPHERE

If you live south of the equator, you live in the Southern Hemisphere. Constellations in the sky seem to move across in a clockwise direction.

It is easier to spot more stars when you are somewhere that doesn't have bright lights.

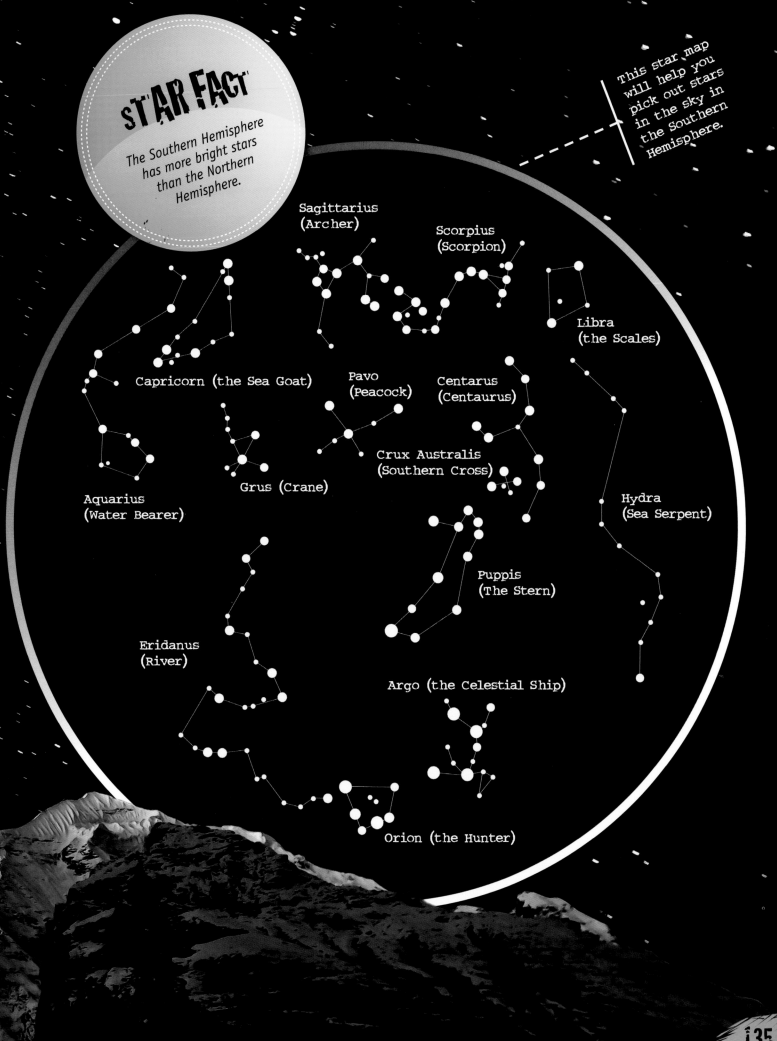

This star map will help you pick out stars in the sky in the Southern Hemisphere.

Sagittarius (Archer)

Scorpius (Scorpion)

Libra (the Scales)

Capricorn (the Sea Goat)

Pavo (Peacock)

Centarus (Centaurus)

Aquarius (Water Bearer)

Grus (Crane)

Crux Australis (Southern Cross)

Hydra (Sea Serpent)

Puppis (The Stern)

Eridanus (River)

Argo (the Celestial Ship)

Orion (the Hunter)

TELESCOPES AND OBSERVATORIES

Without telescopes, astronomers would know very little about the planets, stars, and galaxies. Objects in space are so far away that their light is very faint by the time it reaches Earth. We use telescopes to focus this light and make the object clear for us to see. Some space telescopes are very powerful.

Telescopes can move around to capture the best images.

The right location

A telescope that is used by professional astronomers costs millions of dollars. To get the clearest view, many observatories are positioned on high mountains, above the Earth's weather and away from bright lights. Usually an **observatory** has many telescopes, which may be shared by several countries and universities.

An observatory is a large dome, with telescopes, and workshops inside. Mauna Kea Observatory is on the island of Hawaii. It sits at 13,800 feet

Radio telescopes

The Very Large Array (VLA) is a line of 27 dishes, each measuring 82 feet across. The dishes work together to collect radio waves from distant stars and galaxies. Radio telescopes such as the VLA have mapped the shape of the Milky Way galaxy.

The Very Large Array dishes are about 22 miles apart, arranged in a "Y" shape.

ST'AR FACT

The largest telescopes can see about 100 billion different galaxies. We've not seen them all yet!

TELESCOPES IN THE SKY

Space telescopes
are used to study the Universe
without any disturbance from Earth's
atmosphere. They provide a view of Earth
and other space objects. The first space
telescope, called Uhuru, was launched in 1970.
The Hubble Space telescope is the most famous
telescope used today.

Famous telescope

The Hubble Space Telescope was launched
in 1990 and is one of the greatest
space projects ever. It orbits Earth
and has taken images of many
space objects, including planets
and the farthest galaxies.
Astronomers are using the
Hubble to work out the size
of the Universe.

This special picture is called the Hubble Deep Field. It was made by joining together more than 300 images that were taken by the Hubble. It helped to discover around 1,500 new galaxies.

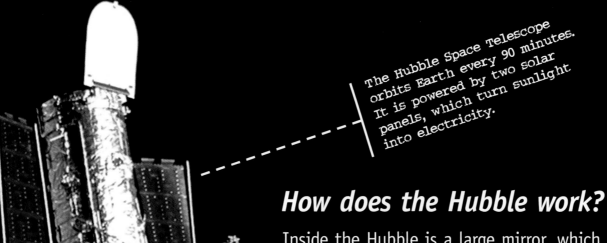

How does the Hubble work?

Inside the Hubble is a large mirror, which supplies light from the space objects that it is studying to five special instruments. It can detect all forms of light, including infrared and ultraviolet. The Hubble has three cameras and sensors that point it to exactly the right area. Astronomers on Earth collect its observations using radio signals.

The Hubble Space Telescope's main mirror measures 94 inches across.

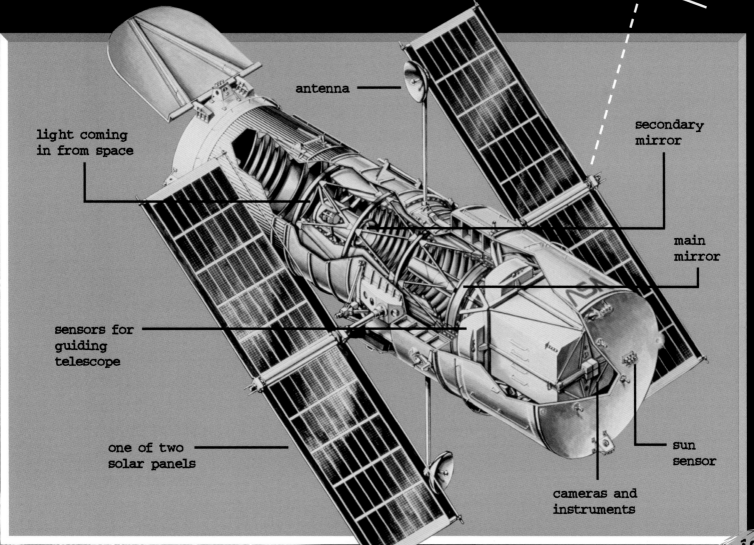

antenna

light coming in from space

secondary mirror

main mirror

sensors for guiding telescope

one of two solar panels

sun sensor

cameras and instruments

GLOSSARY

Asteroid
a rocky object that orbits the Sun. An asteroid is also called a minor planet. It is similar to a planet but much smaller

Astronaut
a person who travels into space. A Russian astronaut is called a cosmonaut

Astronomer
a person who studies astronomy

Astronomy
the study of the Universe

Atmosphere
a layer of gas around the outside of a planet or star

Black hole
an area in space with extremely strong gravity—so strong that nothing can escape, not even light

Cluster
a family of stars or galaxies

Comet
a frozen chunk of ice and dust that orbits the Sun. Gas and dust stream out into space to form its tail

Constellation
one of 88 known star patterns that together cover every part of the sky

Core
the middle of something

Crater
a bowl-shaped hollow on the surface of a planet or moon. Craters form when rocks from space crash into a planet

Crust
the thin, outside layer

Diameter
the largest distance from one side to the other of a circle or sphere

Ellipse
a shape that looks like a squashed circle

Elliptical
shaped like an ellipse

Energy
what is needed to make something move, or to heat something up

Equator
an imaginary line around the middle of a planet

Galaxy
a family of billions of stars

Gas
a gas can flow and move very easily. It will expand to fill a larger space

Gravity
a pulling force that acts between all objects in the Universe

Hemisphere
on a planet, the northern hemisphere lies to the north of the equator and the southern hemisphere lies to the south

Latitude
a measurement that shows
how far north or south of
the equator a place is

Lava
liquid rock that flows
out of a volcano

Lens
a piece of shaped glass or
other transparent material
that is used to focus on
things that are far away

Light
a form of energy that can
travel on its own even
through empty space. Humans
can see visible light

Light year
the distance that light
travels in one year

Liquid
a liquid can flow easily,
but unlike a gas, it does
not expand if moved to
a larger space

Lunar
on the Moon or to do
with the Moon

Magnetic field
a region where
magnetism acts

Mantle
the main layer
of rock that lies
under the crust
of a planet

Matter
matter makes up all
things in the universe,
whether gas, liquid,
or solid. Nearly all
materials can change
between being a gas,
a liquid, or a solid

Meteor
a light in the night sky. It
is caused by a piece of
space dust or rock speeding
into Earth's atmosphere,
where it burns up

Meteorite
a chunk of rock from space
that lands on the surface
of a planet or moon

Milky Way
the galaxy that the Solar
System belongs too

Molten
a liquid. Usually something
that is normally solid,
but has become a liquid
when heated

Moon
a natural body orbiting
a planet. Moons are also
called natural satellites.
The Moon is Earth's
natural satellite

Nebula (plural nebulae)
a cloud of gas or dust
in space

Nucleus
the center of an object

GLOSSARY

Observatory
a place where astronomers use telescopes, or have their offices

Orbit
the path taken by one object around another in space, because the two are attracted together by the pull of gravity

Particle
a very small piece of matter

Planet
a body that orbits the Sun or another star, and does not give out any of its own light

Pole
the poles of a planet or moon are the places where the axis on which it spins meets the surface. There is a north pole and a south pole

Pressure
the strength of the force with which something presses

Radio waves
a form of energy that travels on its own through space and is invisible

Satellite
a small object that orbits something larger. Artificial satellites are manmade objects put into orbit around Earth. Astronomers call the moons that travel around planets "natural satellites"

Solar
to do with the Sun

Solar System
the Sun and everything that orbits around it. The Solar System includes eight major planets and their moons, asteroids, comets, meteors, and dust. They are all held in their orbits by the pull of the Sun's gravity

Solid
a solid cannot flow, and keeps its shape

Space
everywhere beyond Earth's atmosphere

Star
a large ball of hot gas that gives off light and heat

Supernova (plural supernovae)
a massive star that has exploded

Universe
all of space and everything in it

INDEX

Andromeda Galaxy 11, 15
Apollo spacecraft 101, 112–115
Ariel 79
Armstrong, Neil 113, 114, 115
asteroid belt 89
asteroids 12, 13, 43, 65, 69,
    88–89
astronauts 7, 104–105, 108,
    109, 112, 113, 114, 115
astronomy 124–139
Atlas rockets 100

Big Bang 6, 8–9, 25
binoculars 125
black holes 20, 23, 39, 40–41

Callisto 73
carbon dioxide 57, 111
Cassini-Huygens spacecraft 121
Ceres 85, 96
Charon 87
clusters 14–15, 18, 23, 34
Comet Hale-Bopp 92
Comet Shoemaker-Levy 9 93
comet tails 90, 91
comets 43, 54, 90–93, 95
communications satellites 111
constellations 35, 126, 127,
    129, 132, 134
craters 54, 57, 64, 65, 77, 83

Dactyl 88
Dawn spacecraft 96
days and nights 62, 64, 68, 72,
    78, 87
Deimos 69
Doppler effect 25
dwarf planets 84–87
dwarf stars 36, 37, 38, 39

Earth 10, 32, 42, 43, 46,
    52, 53, 58–63, 67, 91
eclipses 13, 32–33, 67
Einstein, Albert 12–13
elliptical galaxies 16
energy 8, 9, 30, 38
Eris 63, 84, 85
Europa 73

Gagarin, Yuri 99
galaxies 6, 7, 8, 9, 11,
    14–23, 25, 138
Galilean moons 73, 74
Galileo Galilei 74–75
Galileo spacecraft 121
Ganymede 73
gas giants 70–73, 76–79,
    82–83, 120–121
giant and supergiant stars 36,
    37, 38, 39
GPS satellites 111
gravity 9, 13, 14, 17, 18,
    19, 23, 34, 40, 41, 42,
    43, 44, 50–51, 55, 64,
    69, 77, 99, 117
Great Dark Spot 83
Great Red Spot 73

Halley's Comet 92, 95
Haumea 85
Helios spacecraft 122
helium 8, 9, 70
Herschel, William 80–81
Hipparchus 128
Hubble, Edwin 24–25
Hubble Deep Field 138
Hubble Space Telescope
    24, 138–139
Hubble's law 25
Hydra 87
hydrogen 8, 9, 70

ice 69, 71, 77, 83, 86, 87
International Space Station
    102, 106, 108–109
Io 73
irregular galaxies 17

Jupiter 43, 46, 70, 72–73,
    74, 89, 120, 121

Kepler, Johannes 48–49
Kuiper Belt 84, 85

landers 97, 118
Large Magellanic Cloud 15
laws of motion 51
life zone 59
light years 11
Luna spacecraft 99
lunar eclipses 67

Magellan spacecraft 117
magnitude system 128
Makemake 85
Mars 43, 46, 52, 53, 68–69,
    96, 116, 117, 118–119
matter 8
Mercury 43, 46, 47, 52, 53,
    54–55, 116, 117
meteor showers 95
meteors and meteorites 64,
    94–95
methane 70, 82, 86
Milky Way 6, 14, 15, 22–23,
    40, 137
Mir space station 107
Moon 7, 10, 32, 33, 52, 64–
    67, 99

INDEX

Moon landings 112–115
Moon, phases of the 66–67
moons 45, 69, 73, 74, 77, 79,
    83, 87, 88

nebulae 34, 35
Neptune 11, 43, 46, 70, 71,
    82–83, 120
Nereid 83
Newton, Isaac 50–51
nitrogen 77, 86
Nix 87
Northern Hemisphere 132–133

Oberon 79
observatories 136–137
Olympus Mons 69
orbiters 68, 97, 118
orbits 22, 46, 47, 48, 54, 62,
    63, 66, 82, 87
Orion 35, 133
oxygen 58, 105

Pathfinder spacecraft 119
Phobos 69
planets 42–43, 45, 46–49,
    52–63, 68–73, 76–79,
    82–87, 116–121
planisphere 131
Pluto 84, 85, 86–87
plutoids 85
prominences 31

quasars 20, 21

radio galaxies 20
radio waves 20, 22
relativity, theories of 12, 13
rings 71, 76, 77, 78, 79, 82
rockets 97, 100–101, 103
rocky planets 42, 43, 52–63,
    68–69, 116–119
rovers 119

Salyut space station 106
satellites 97, 98, 110–111
Saturn 43, 46, 70, 71, 76–77,
    120, 121
Saturn rockets 101, 113
seasons 63
Sirius 27
Skylab space station 107
Solar and Heliospheric
    Observatory (SOHO) 123
solar eclipses 13, 32–33
solar flares 30
Solar System 7, 26, 28,
    42–47, 71
solar winds 29, 90, 123
Southern Hemisphere 81,
    134–135
Soyuz spacecraft 106
space exploration 96–123
space probes 97, 116, 120,
    121, 122
space shuttles 102–103
space stations 102, 106–109
space tourism 96
spacesuits 105
speed of light 10–11
spiral galaxies 9, 15, 17, 23
Sputnik 98, 99
star maps 126, 130–131

stars 6, 7, 9, 18,
    19, 22, 23,
    26–39, 124,
    125, 126–135
storms 68, 76
Sun 7, 10, 11, 13, 22, 26,
    28–33, 36, 37, 38, 42,
    43, 44, 46–47, 48, 63,
    122–123, 127
sunspots 31
supernovae 39

telescopes 20, 24, 74, 80,
    124, 125, 136–139
temperatures 28, 29, 30,
    35, 36, 37, 55, 57, 59,
    61, 78, 83, 122
Tiangong-1 space station 106
Titan 77, 121
Titania 79
Trans-Neptunian Objects 84
Triton 83

Ulysses spacecraft 123
Universe 6, 8, 25
Uranus 43, 46, 70, 71, 78–79,
    80, 81, 120

Valles Marineris 68
Venus 43, 46, 52, 53, 56–57,
    74, 116, 117, 120
Very Large Array (VLA) 20, 137
Viking spacecraft 118
volcanoes 53, 54,
    57, 60, 69
Voyager spacecraft 120

weather satellites 110

zodiac 126, 127